Joyworks

The Story of Marquette Electronics
and Two Lucky Entrepreneurs

written by
Michael J. Cudahy

MILWAUKEE COUNTY HISTORICAL SOCIETY

Editorial services by Sarah Kimball
Designed by Kate Hawley
Cover photo by Scott Paulus
All other images from the Michael J. Cudahy archive

When Irish eyes aren't smiling... by Meg Kissinger
Milwaukee Journal Sentinel, June 23, 2000
© 2000 Journal Sentinel Inc., reproduced with permission.

Cudahy uses trust, perks to motivate Marquette workers by Avrum D. Lank
Milwaukee Sentinel, December 13, 1988
© 1988 Journal Sentinel Inc., reproduced with permission.

Hitler on America by John Cudahy
© 1941 Time Inc. reprinted by permission.

ISBN 0-938076-17-5

Printed in the United States of America.

To my children: Susie, Julie, Patrick,
Mary, and Michael as well as my
adopted son, Jason. Also to my stepchildren:
Kathy, George, Carrie, Peter, and Amy.

Lest I forget my many wives:
Mary Lee, Audrie, Nancy, and,
of course, Lisa, with whom I still live.

Also, my wonderful sister Mary
and my tolerant mother and father.

And, of course, to my partner
Warren Cozzens and his family.

To all of the dedicated employees I have
worked with throughout all the years.

My love for all of these wonderful people is,
and will be, forever. Without these great – if
often convoluted – relationships and my Irish
luck, I'm sure I would not have been able to
survive, much less prosper as I have.

List of Chapters

Foreword

In this remarkable success story, Michael Cudahy's Irish wit, compassion, and fascinating background come shining through page after page. This is a must-read tale of how an entrepreneur with little formal education went against the grain and used his intellect, wit, and charm to make his dream a reality.

Mike had a vision. The resulting medical advances improved the quality of health care worldwide, as well as the quality of life in communities across America, especially in our home state of Wisconsin. He also had a common sense vision of how to treat people, build loyalty, and change the way businesses are run.

Marquette Electronics was the right company at the right time – and Mike and his partner Warren were the right people at the right time – to help usher in a new era in medical care. Mike and his team were at the forefront.

We live in an era of unprecedented opportunity to help families live healthier, more productive lives by taking advantage of the stunning advances in medical technology. The advances we have seen during our lifetime, with more forthcoming every day, are a major part of that opportunity.

Mike is an American entrepreneur in the truest sense of the word. From his limitless curiosity as a young boy – helped along by infinitely patient parents – to his extraordinary generosity as a man, Mike has led an exceptional life.

As Mike's story unfolds, he and his family meet world leaders, jazz greats, business giants. Numerous people from the world over count themselves fortunate to have met him. I am happy to say I am one of those people and am blessed to have him as a special friend.

Mike's inventiveness, generosity, and kindness have made our state, country, and world a much better place — something that can be said about few people. *Joyworks* gives the inside story on how that inventive and sometimes mischievous mind translated into a better world.

Tommy G. Thompson
Former Governor of Wisconsin

Preface

The name Cudahy is one of the most prominent in Wisconsin. In this lively and interesting volume, Mike Cudahy follows in the footsteps of his father and grandfather. Both John and Patrick Cudahy were innovative and successful in their chosen lines of work; both understood the importance of writing about their lives; and both knew how to tell a good tale.

Patrick Cudahy (1849-1919), the family patriarch and founder of the meatpacking firm that bears his name, published a delightful autobiography in 1912. His son John (1887-1943), a successful businessman, lawyer, and diplomat, was the author of four books. Like his father, he wrote on subjects that he had experienced directly, ranging from the campaign of U.S. 85th Division in Russia after World War I and his diplomatic experience in Europe to hunting and photographic expeditions in Africa and Lower California.

Mike Cudahy continues in that same tradition. His volume bears similarities in spirit to his grandfather's book in that it is both witty and candid. In telling the story of his life and of his successful business, Marquette Electronics, Cudahy avoids the usual traps of business history. Too often these types of accounts get mired in statistics: how many people were employed; what was the volume of sales; what was the rate of growth. Numbers become more important than people. As Mike Cudahy notes, the success of Marquette Electronics was more about people and culture than business plans. *Joyworks* keeps good stories and people at the center.

Modern business generates mountains of paper, yet frank accounts written by insiders are more useful to future historians than the officially sanctioned (and sanitized) reports written by corporate public relations departments. The publication of *Joyworks* continues the Cudahy saga. Patrick and John Cudahy would have been proud.

Michael E. Stevens
State Historian of Wisconsin
Wisconsin Historical Society

Introduction

This is not one of those *how to* management books you see around the stores. There must be thousands out there these days, written by people with much more writing skill and knowledge than I. Besides, I think most people are rather of sick of them. Not that I haven't read a few, and not that some aren't valuable reading, like Townsend and Peter and don't forget Deming. Townsend's and Peter's books have influenced me a lot and I'm a total worshiper of Deming.

This is a book on *how it happened*. How Marquette Electronics started with one employee and $15,000 back in 1964 and grew to be a world leader in the field of medical electronics and some of the *rules* I used to guide it through the years. It's not quite the same as Dave Packard's book, *The HP Way*. You'll see that *Joyworks* has turned out to be more of a comedy.

Quite a few people have asked me to write the story of Marquette. I'm not totally clear as to just why they ask. I'm flattered, but I think it might be a bit of morbid curiosity. "How in hell could this high school dropout [well, almost] con his way into developing one of the major medical electronic manufacturing companies with almost zero money," is what goes through their minds, I suspect.

People who know the story, superficially, don't always believe it so I guess it's appropriate to set the record straight if I can.

To start with, many think I inherited gobs of money from my rich and successful meatpacking grandfather. He's the guy who quit school at thirteen and eventually started his own company that employed some six thousand people at its peak.

What they don't realize about my grandfather was that he practiced primogeniture. That's where the oldest son takes everything. The inheritance from Grandpa Pat, the business, everything went to the oldest son, and in this case it was *not* my dear father – rather, his brother, Michael, my Uncle Mike.

That's what they did in the old days, especially the Irish. The oldest son, right from early childhood, would be trained to take over the business if the old man was successful in creating an enterprise of any magnitude. Never mind if the poor guy *wanted* to do the job. Take the responsibility or else, and that was that. In those days the girls in the family were not even considered.

My grandfather's life is a story in itself. There aren't enough pages here to even start to describe this phenomenal man from the old sod, but a few words might help the reader understand the strange and wonderful family I come from.

Michael J. Cudahy

Patrick Cudahy

CHAPTER 1

THE EARLY, EARLY DAYS

I n 1849, Ireland was on its knees with the worst potato famine the country had ever seen. People were dying by the thousands, and the Shaw family of Callan, County Kilkenny, decided that they had had enough. Shaw's daughter was married to Patrick Cuddihy (old world spelling), my great-grandfather.

Fortunately, old man Shaw (said to be a relative of George Bernard Shaw) had a pottery business in Callan and found someone (believe it or not in such bad times) to buy the business for three hundred pounds. That might translate back in those days to about $1200, so the family, numbering nine in all, was able to book passage on a sailing vessel called the *Goodwin*, steerage class, no doubt.

These vessels were known as *coffin ships* because of their miserable, cramped quarters and the high fatality rate in crossing the Atlantic.

My grandfather, Patrick, was one of the youngest travelers at the age three months. It is written that the voyage took two or three months, and I'm sure there was no supply of Gerber's baby food for little Pat.

When the family arrived in Boston, it didn't take the Shaws and the Cuddihys long to decide to push on to *the land of milk and honey* as the Midwest was then called. The Irish were not welcome in Boston at that moment of American history. Signs posted in business places said, "Help wanted. Irish need not apply," an early bit of discrimination in America, perhaps brought on by the British in Boston or the disorderly conduct of some of the many Irish.

And so this wandering family of immigrants, the Shaws and the Cuddihys, apparently found their way to Milwaukee, Wisconsin, where, it is said, they had friends. My grandfather, Patrick, then quit school at the age of thirteen and went to work for a local meat-packing company by the name of Plankinton, later affiliated with the well-known Armour Packing Company of Chicago. In his autobiography, Patrick said that people always need the basics, such as meat, so this would be a very good industry to pursue. Young as he was, the family was apparently destitute, so there was no choice but work.

He ascended at Plankinton, even as a young boy, by demonstrating a keen eye for innovative ways to improve the company. It seems hard to imagine how this *kid* made such an impression on management, but the records show that Patrick advanced steadily, especially in his early twenties. So much is written these days about genes and how they affect one's career that it seems logical Patrick had something in his blood other than an iron will to keep the home fires burning.

In any case, meatpacker John Plankinton was so impressed with Patrick at age twenty-five that he sold the young man one-sixth of his business. And in 1888, Plankinton retired and sold his remaining interest to Patrick, with financial help from Pat's brothers, Michael, Edward, and John. The company was then renamed Cudahy Brothers Company.

Brothers Michael and Edward worked for Armour in Chicago. Later these two entrepreneurs formed the Cudahy Packing Company, which became one of the largest firms of its kind in the nation. Was it genes, Irish luck, determination or a blend of all – who knows?

Patrick's brother, John, was a bit of a roué, however. He speculated on the grain market and amassed a fortune, but later lost it all during the depression of the 1880s. A Colonel McNeery, writing for the *Chicago Post* in 1883, said of John Cudahy:

I heerd th' other day that me old friend Jawn Cuddy was on th' r-run. They've been callin' him Coodyhie since he got all th' money, but I'll bet it will be Cuddy now he's broke. "Well" I says, "I'll go down to the boord iv thrade an' see him go broke," I says. An' with that I up to th' boord and sits in th' gallery an' puts me feet on th' railin'. The crazy lads was dancin' around on th' flure an' shakin' their fists under each others' noses. Then a man turns to me an' yells "Boots," and ivery man on the th' flure stops fightin' an' points to me an' says "Boots" an though I was wearing paytent lither ties, I ups an' leaves them. An' I hear no more about Cuddy till a man come into th' place an' says, "Cuddy's gawn."

Cuddihy, Cuddy, Coodhie, Cudahy? It is suspected that Michael, my grandfather's oldest brother, suggested the spelling of the name be changed to Cudahy somewhere in the 1860s because he thought it would be easier to pronounce.

Meanwhile, in New York, another branch of the Cuddihys, also from Callan in County Kilkenny, migrated at about the same time as Grandfather Patrick and his family. Young Robert J. Cuddihy worked his way up in the publishing industry, eventually heading a very popular magazine called the *Literary Digest*.

According to a book entitled *Real Lace* by Stephen Birmingham, grandfather and R.J., as he was called, met in New York sometime in the 1890s and discussed the spelling of the name,

R.J. maintaining the old world *Cuddihy* and Patrick, with the new spelling *Cudahy*. R.J. was to have said:

> Both our families came over in the forties from the same county, Kilkenny, and the same town. Only there was difference in our two forebears – *ours* knew how to spell.

Grandpa Cudahy passed away in 1919 and since I was born in 1924, I never had the chance to meet this wonderful man. Pity, for he must have been one character and a half. And since he practiced primogeniture, where the eldest son becomes chief heir, Uncle Mike landed in the driver's seat at the Cudahy Brothers meat packing plant, and my father John went the other way.

First, to law school in Madison. After graduation, he served as

John C. Cudahy

the chief attorney for the company, but I think he found it a crashing bore, and I'm not sure how well he and Uncle Mike hit it off. Big brother was the boss. Father dabbled a bit in the outside practice of law, but his first love was politics.

My parents were married in 1913 and started their life together in a modest apartment on Ogden Avenue in Milwaukee, living a sort of upper middle class life. Primogeniture probably didn't leave them too much, yet I'm sure, they were not destitute. After all, Grandpa Cudahy was one of the most successful businessmen in Wisconsin at that time, and I'll bet the seven members of his family other than Uncle Mike (six daughters plus my father) were not neglected.

My fourth ex-wife, Lisa, with whom I live, has listened to my chitchat of all this and about my mother and father to the point of nausea. One day she remarked, "It sounds like your parents were a totally odd couple, completely at opposite ends of the scale."

She might be right. My father was tall, dashing and outspoken. He loved to read books and philosophize on world politics and give speeches and be in the limelight. I suspect his long-range master objective was to become the president of these United States of America. He was one hell of an impressive guy, and if it weren't for his untimely death in 1943 at the age of fifty-six, he just might have made it.

My mother, on the other hand, was small, pretty and modest and hated the limelight in every way. When my father became well known, the press followed his every move, and my mother used to run and hide from it all. There are, literally, thousands of photographs of my father — scarcely any of my mother.

Yet they lived, I think, in reasonable happiness

Katharine Reed Cudahy

Mike Cudahy and his sister, Mary

although my father was, to a large extent, an absentee husband and father. His career was very important to him and took him to many, many far-off lands. We were sometimes along, but more often than not, left at home because of traveling complications.

Father started his political career running for various offices — alderman, state senator and lieutenant governor, but was not too successful. During this time, however, he made the acquaintance of one James Farley. Born in 1888, Farley was the politician's politician and largely responsible for Franklin D. Roosevelt's stunning presidential victory in 1932. He also headed the Democratic National Committee and, for some reason, took a liking to my father.

Farley, then Postmaster General under Roosevelt, changed the course of my father's life by persuading Roosevelt to appoint father U.S. ambassador to Poland. I don't think the old man knew a bloody thing about being an ambassador, but that was his job for the next four years, and a big one it was.

So off we went to Warsaw, a godforsaken place in those days — dark and cold and still devastated from World War I. My mother was a wreck, I can remember. She was having a fit about what was going to happen to us kids. What about schools? What about

John Cudahy and James Farley

doctors? What about friends? Whatever would happen to my sister and me in this strange and foreign place so far from home? My mother was the greatest worrier on God's green earth, and this gave her a lot to worry about.

But we somehow survived. The parents found a doctor in case we got sick. As to friends, that was another matter. My sister might have managed to find a few, but it wasn't easy, what with the language barrier. I took up with the embassy staff, those who could speak English. As to whether they took up with me, I do not know, since they had little choice. After all, I was the ambassador's son, nine years old and probably a first class brat.

There were moments when I wondered what I was going to do with myself. At nine, I was full of energy – restless, and had almost an obsession for anything mechanical or electrical. I searched for things to take apart, hoping I would find out how they worked. Perhaps this quirk was what propelled me into the field of electronics many years later.

I found little around the embassy in Warsaw to disassemble and investigate except a large console, 78 RPM Victrola record player. Not very interesting, I thought, except what strange device amplified the sound? The turntable motor was simple – a big spring, some gears, a winder. But what made the sound get from the disc into the room? I was sure there was some magic box inside that did all this and I had to find out. The only way was to take the thing apart.

I asked a guy on the staff by the name of Stanislas for a screwdriver and he said, "Vut do you vant dat for?" and I really didn't want to say, for fear he'd say "no." So I told him there was a squeak in the Victrola, and I knew just where to put the oil if only I had some oil and a screwdriver. To my delight, he said, "I get you some tools and oil and I give you help."

What I wished he hadn't said was "I give you help." That meant he'd come along with the deal, and I wanted to be alone to do my snooping. I tried to act like some budding American genius with my comments on what should be done. I don't think he believed a word I said, however. But just then he was called to the phone, and I was alone.

"Rip into it, Mike, and find out once and for all what goes on in this machine, and do it quickly before this guy gets back," I said to myself. Four screws on the top panel and out came the whole turntable assembly. I lifted it up until the underside was exposed. And much to my dismay, there was nothing under the panel at all – only the windup motor and empty space that seemed to go to the grill in front.

How can this be, I thought? How does the sound get out into the room? Wasn't there supposed to be an amplifier in there somewhere? Unfortunately, Stanislas returned from his phone call and said, "Vell, vell, how are ve doing here, young man? My, you got da ting half apart! Look like you and I are going to have to try to poot it back together before vee loose some parts. I guess vee just going to have to live wit da squeak, OK?"

What a bummer! I was burning up with curiosity, and now I might never get back inside this machine and find out what made it work.

I demonstrated my brattiness in another big way in Poland, so big it almost caused an international incident between Poland and the United States. It was the night of a very official dinner party for some fifty distinguished guests, many of world prominence.

That week, someone had shown me a clever trick. First, you take a cigarette and cut it in three equal parts with a good, sharp razor blade. Now, replace the center section with a carefully rolled cylinder of Kleenex and Duco cement, a very flammable mixture, indeed. Then roll all three pieces together with some roll-your-own paper, and you have a real jewel of a trick for some unsuspecting victim.

It took hundreds of tries to finally get one of these *bombs* to look almost perfect, but I finally succeeded and since the old man had a great sense of humor, I proudly brought my great *invention* to him to see what he'd think.

He was impressed, all right, but cautioned that some of the dignitaries might not think the idea was that funny, especially if they were to singe their mustache or something even worse. No, he said, the only guy I could try it on was one of the American

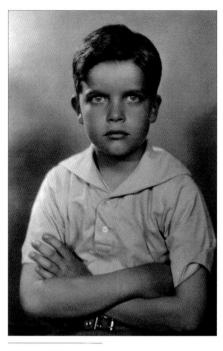

Mike Cudahy at age 8

staff members, a guy also with a sense of humor we called "Eggy."

In total defiance of these sage instructions, however, I carefully placed the bomb in the very best silver cigarette box on a table in the living room where, surely, one of the distinguished guests would fall victim.

And *fall victim* one did. Sometime during that evening, long after I was told to go to bed, my father had a rather delicate diplomatic conversation with a gentleman who had been seeking out the ambassador at the dinner table, an extensively decorated and important Polish general. Their subject was the decaying Polish-American relations, and just as the conversation reached a critical moment, the good general reached for a smoke. As the tension grew he took a deep puff and – poof! The general was with blackened face and singed mustache.

"Outrageous!" the general blurted, and stormed out the door.

The press carried an article, which more or less explained the incident as follows:

Milwaukee, where "Peck's Bad Boy" was born, may now boast of young "Mike" Cudahy, 9-year old son of

John Cudahy, United States ambassador to Poland, according to a story in The New Yorker.

Mike has been guilty, according to reports, of a number of "unfriendly acts," as the diplomatic language refers to such things. One of these concerns the placing of some explosive cigarettes in various cigarette boxes about the embassy drawing room prior to a reception which his father was giving for members of the diplomatic corps and other notables about Warsaw.

One of these innocent looking cigarettes, it is related, was picked up by a "grim Polish general" who put great store by his military pose. It exploded while he was trying to impress some elegant and brilliant woman with his grasp on something or other of worldwide importance. The general was so mad that he departed without bidding adieu to his host or hostess – a real diplomatic incident.

My father summoned me to his office the next morning, and I had a pretty good idea what he wanted me for. He exclaimed about the seriousness of the matter, saying that Polish-American relations might be in jeopardy because of my foolishness. I shook with fear as he reviewed the events of the night before, but just then I could see a little smirk on his reddened face, and I realized it was all over. Suddenly we were both in hysterics, and he blurted, "Now get out of here, and don't ever do a thing like that again, do you hear?"

Believe me, I never did.

Harrison Reed

CHAPTER 2

TINKER, TINKER, PUTTER, PUTTER

Back in the States for awhile when I was about ten, Mother used to require that my sister and I (there were only two of us) visit Grandma and Grandpa Reed (the other side of the family) almost every Sunday. We'd show up and make some small talk with Grandma, and then I'd high-tail it out to the garage where Grandpa had his shop, an awesome place to me, with all sorts of wonderful tools and machines.

Harrison Reed was a *putterer* of the first order and when he wasn't at his hardware and fishing tackle shop, he'd spend his time making amazing things for the house or his store or a friend – or just for the fun of it. He was one of those guys who had infinite patience and could fix anything on earth. He always whistled *Roaming in the Gloaming* when he worked. As a young lad, I was absolutely fascinated with him. Those Sundays I spent in his garage, I suspect, influenced me as much as anything in my whole life, and perhaps without him I might never have succeeded in the career I pursued.

"Whatcha building now, Grandpa?"

"Door knob. Old one had a bad sleeve."

"Whaterya makin' it outa, Grandpa?"

"Brass, solid brass. Only way to make it so's it'll last."

"Can I help? Can I run the lathe?"

"Nope. Might hurt yourself. Oh well, step up here and be careful – I'll show ya' how. I said careful there!"

Many a Sunday was spent with Grandpa during that part of my life when I was so impressionable, and I often wonder about the golden opportunity so many of us have had to influence a young lad in such a positive way. The Reed side of the family was mechanical, especially Grandpa. My father couldn't change a fuse if his life depended on it.

About that time in my life, I started to tinker with wires and things like that. I remember there was a picture in our house that I liked a lot but it was in a dark part of the hallway, upstairs.

One day, out of the blue, I set out to wire up a light over this picture and, although it doesn't sound like a big project, I had to start from scratch with nothing but a flashlight bulb, a battery, and some wire. A short time before that I stuck a hairpin in one of the outlets in the house and blew a fuse so my mother wisely forbade me from tinkering with anything hooked to the power line.

Carefully, I rigged this silly little bulb in some sort of socket over the picture and ran wires down the corner of the wall to a battery, and things were looking pretty good. Of course, the battery soon ran down, and I can remember pondering and pondering about this huge problem in my life.

Eventually, however, I thought of a possible solution – an *invention* in my mind. With a lot of tinkering, I managed to put part of the wiring under the rug in such a way that the light would only go on when you stepped on the carpet in front of the picture.

My mother was so proud of her budding young genius that every single visitor who came to our house from that day on had to come and see Michael's invention. And the boost I got from my admirers set me off onto a passion for electricity that never ended.

I was intrigued with mechanical things, too. My father had a beautiful gold pocket watch his father gave him when my parents were married in 1913. I always had my eye on this wonderful machine, not to take it, God knows, but just to be alone with it for awhile and see how the tiny little chime inside worked.

One day, the opportunity came. My father had gone to town and apparently forgotten it, for there it was on his dresser, just waiting for me.

I brought it to my room where I had a fair number of tools – not exactly watchmaker's tools, just ordinary tools. I was lucky in not encountering my mother or my nosy sister on the way, and I even locked the door when I got to my room to make sure I wouldn't have some inquisitive visitor during such a delicate investigation.

Unfortunately, the back of the watch was not easy to remove. The tools were so big and clumsy that I had to resort to a Swiss pocketknife my grandfather had given me the year before.

Carefully (I thought) I pried around the edge until, bing, the back fell to the floor and rolled halfway across the room. Panicked, I froze, wondering what else had fallen off and how to put this delicate treasure down so I could retrieve the back.

I managed, however, and now I felt compelled to go on. I could see no trace of the chime or much of anything for that matter. There were many tiny screws on the back panel, now exposed, and I decided they had to come out if I were to get anywhere.

The Swiss knife was horribly clumsy, but somehow, one by one, I removed all eight screws and pried up the back panel. Suddenly there were parts flying everywhere. Springs, tiny gears, wheels, the chime, and every other part known to watchmaking.

The next three hours were sheer hell as I tried every way possible to arrange the parts in some logical order and to replace the back plate. I could hear my heart pounding as I assessed the situation, but there was no way to even guess where all the parts should go, much less how to get them there. No hope whatsoever, and it looked like I was going to have to confess or think up the biggest lie a boy of ten had ever told.

And now my mother was at the door. "Michael, is that you? Are you in there? Why is your door locked? Let me in!"

"Just a minute – I'll be right there. I'll open up in a second."

"Whatever are you up to in here, Michael? And why did you have the door locked?"

"Uh, I'm having a little trouble with Date's watch (Date was the name my sister and I bestowed on poor old Dad), and I didn't

Date's Watch

want to be disturbed. I was trying to be a good guy and fix it for him 'cause he said it was running fast, but I slipped and a few pieces fell out."

"Oh, my!" said my dear mother as she looked down at the pieces strewn about the floor. "Oh, my! Your father will be furious. We'd better get it down to Mr. Neverman's shop right away before he finds out."

My mother covered for me, as she always did, and my father never did find out. Mr. Neverman, a jeweler friend of the family, somehow got it all back together again. He also gave me an old watch he had lying around his shop saying, "Now, you tinker with this one from now on, Michael."

I eventually learned to take the old *junker* apart and put it back together, but only after several more visits with kindly old Mr. Neverman.

My mother had a very close friend by the name of Olive Fuller. She was a chubby, jolly lady who had recently been widowed. In the lonely days when my father was still in Poland, Olive would come over to see Mother, and they'd knit and talk together on the screen porch for hours, discussing everything under the sun.

One night we were all invited to Olive's for dinner, and I was positive I was going to be bored out of my mind. I figured my sister, Mary, was probably OK because she could make all that dumb small talk with the ladies and bring her own knitting. But I was really in for a bad evening, or so I thought, and the dinner was about as slow and dull as I expected.

Little did I know that I was in for the surprise of my life.

Over dinner, Olive was complaining about how her radio was not working, and my dear mother suddenly piped up with, "Well, Michael, here, is just a little genius about electricity and radios and all sorts of mechanical things. I bet he might be able to fix it for you, Olive."

To my surprise, Olive said, "Oh, wonderful," and showed me a little old-time radio down the hall. It was one of those all-wooden table models with three tuning knobs on the front panel and a big *morning glory* speaker on the top.

"Hot dog!" I mumbled to myself. "What if I can really fix it?" I told Olive, with a tone of authority, that I'd need a screwdriver and a pliers, at least for a start. "Will you need some extra light on the subject?" she asked. I said that might help, as if I knew what I might be looking for.

Then it happened. My best recollection is that I took the cover off and looked around for about half an hour in absolute amazement, wiggling a few wires here and there and for some reason, I then decided to put it back together and see what might happen.

To my utter dismay, the old set started to play! Quick, what did I do and what will I say to them? I could hear them hurrying down the hall as the music blared from the horn.

"Michael, you've fixed it. You've fixed my poor little radio. Why, Katharine, the boy's a genius! Listen to how wonderful it sounds."

I beamed. My mother beamed. "See, Olive, I told you he was a little genius. I just knew he'd fix it."

My sister covered her face with her hands and mumbled, "Euh buoy," as I unashamedly took full credit.

Another idea that intrigued me at that age was to find a way to turn on a radio with an alarm clock. The year was 1939, and although I may have read of such a thing somewhere, alarm clock radios, per se, weren't available off the shelf in stores, or so I thought. You could buy small, table-top radios, and electric clocks had been around for some time, but not these two together.

I thought that my favorite radio program would be a cool way to wake up for school in the morning, so I went to work on the "GE Telecron" clock my mother gave me, hoping that I'd be better at getting to school in the morning.

I was to have more luck this time than with my father's gold watch fiasco. By now I had some practice taking apart the old watch Mr. Neverman so kindly gave me after I wrecked my father's.

I remember constructing a crude, dangerous electrical contact on the arm that released the bell inside this clock, splitting the power line to the radio so that electricity would come through the clock only at the "alarm" time. I was almost unable to sleep that first night in anticipation of the blare of music that would be coming my way the next morning, but the music never came. It took another three or four tries, but finally it worked, and I spent that entire day at school bragging about my invention until some lousy kid came up and said, "You didn't invent that, Cudahy. My dad gave me one for Christmas that he bought from a catalog."

Crushed, I recovered from my bruised ego and went on to think up more novel ways to be awakened. Besides, I was getting used to the sweet music from my alarm clock radio and occasionally slept through it all. There had to be a more spectacular way to greet the day besides music, I reasoned. Suddenly another brainstorm came to mind.

If I installed that quarter horsepower electric motor that Grandpa gave me at the foot of my bed, I could use it to zip off the bed covers when the alarm went off. Wow, how spectacular, and just let some kid try and tell me he had a gadget like that!

Winding twine around the shaft of the motor, I guided it over the foot board and branched it to each side of the bed where I secured the bed covers with alligator clips. The motor would then start, the string would wind up, and the bed covers would come zipping off – I hoped.

I constructed a very special lever switch at the foot of the bed from my Erector Set (a mechanical, do-it-yourself kit that lots of young boys had in those days). A knot in the string would then trip my switch at the appropriate time and turn the motor off.

All looked well as I made final checks and went to bed. The night dragged on with anticipation of the great *rip-off* that was to occur the next morning. I finally drifted off and promptly at seven o'clock the motor began to spin and the bed covers began to vanish. Unfortunately, however, the motor continued its relentless whir because the knot in the string had slipped past the lever switch. The sheets, the blanket, and all other covers were suddenly spinning in a ball, ripping and skidding across the room, totally out of control, destroying themselves while I watched, helplessly.

My mother again forgave me and replaced the linens. I had to agree to discontinue these experiments.

Around this same age, my mother decided that I was also a genius in music with a fantastic ear, so I should take piano lessons and be famous when I grew up.

She found this lady by the name of Mrs. Graham, who would come to the house every week and teach me the scales while the other kids in the neighborhood were out playing baseball. Not only did I want to be out there with the guys, but Mrs. Graham sat right next to me on the piano bench and had bad breath.

It took a long time to get back to some interest in the piano after that, and I suspect I could have made some sort of a career in music, but I didn't. I've always had a love for jazz, however, and when I was about fifteen, I used to try and play boogie-woogie on the piano.

One night I managed to get into a local joint, called the Blatz Palm Garden, on Water Street in Milwaukee (carding was not done in those days, but if you looked too young you could be asked to leave).

Jazz legend Fats Waller was playing the piano and singing – and I was hypnotized. I got a chance to introduce myself, and he was totally gracious and encouraged me to learn to play.

In the years that followed, many jazz artists came to Milwaukee, and I always found a way to get to see them even if it meant I had to lie about my age to get in. Louis Jordan with Five Guys Named Mo, Count Basie, Coleman Hawkins, Art Tatum, Nat King Cole, and a whole host of others. I got to know each one personally.

In 1937 we moved to Ireland because my father had a new appointment there as our U.S. Minister (same thing as ambassador, but Ireland was not yet technically a nation), and I attended a school called Belvedere College in downtown Dublin. A so-called college over there was really not college as we know it but more or less equivalent to our high schools. At 13, I was an equivalent, more or less, to a freshman.

But being the ambassador's son, I suspect, I was somewhat of a curiosity at school. As such, I didn't develop many take-home friends that I can remember. My only young buddy was the chauffeur's son, Sean, who was a year younger. He and I forever searched for evil things to do around the embassy in the typical style of young boys. I was the older and so the leader, and I can remember telling him everything I knew about sex, which was almost nothing.

I did have one other buddy who was a lot older, 25 or 30, named Paddy McCormack. This guy was a friend of one of the staff members at the embassy and was suggested as a companion for me because he was a ham radio operator. Just down my alley, they thought, and they were right. I think the old man felt Paddy might keep me from cabin fever and, believe me, he did. Paddy would come over every week or so, and we'd rap for hours in a language all our own – radio.

Paddy taught me many fundamentals of this fascinating subject, and I absorbed it all like a sponge. Between visits, I read books like *The Amateur Radio Relay League* and *Beginner's Radio*, soon developing a passion to become an amateur radio operator like Paddy. He had told me he could talk to other amateurs all over the world on his rig at home and the more he told me, the more I became hypnotized at the possibility.

Soon I realized my dream. I got some sort of sharing license with Paddy. EI5P was the call sign, and now came the job of building a transmitter and converting an old Philco console radio to operate on the ham radio bands. Buying equipment for this sort of thing, by the way, was not possible in those days. Radio transmitters had to be built from basic parts.

This sort of tinkering had its dangers. A good vacuum tube transmitter required as much as 750 volts or more of DC current to make it perform properly for talking to people in far away places, which was where all the fun was. One day I was carefully tuning a transmitter I was building and, somehow or another, I leaned over, relaxing my arm on the *hot* side of the high voltage. Before I knew what had happened, I found myself halfway across the room from an electrical shock I'll never forget.

Most of the time, however, Paddy was there to keep me safe and was at my side when I successfully talked by short wave with some stranger in England. It's a good thing Paddy was around because I might have jumped through the window with excitement. I was ecstatic – I was thrilled beyond my wildest dreams. I had arrived at my life's work, no question about it.

From that moment on, school, family, sports, friends, even food and sleep were of secondary importance. Only Paddy McCormack and my radio rig meant anything in my life. I must have spent thousands of hours reading, testing, calling "CQ" over the airwaves, the universal call for anyone, anywhere to answer. At one time, I succeeded in reaching and talking to a fellow in Havana, Cuba. When I think of it, I still get goose bumps.

Somewhere during this wonderful time in my life, I learned to test my transmitting equipment for what are called *standing waves*. These were the waves on the transmission line leading up to the outdoor antenna, which deprived the signal of strength and thus affected the distance that I could reach with my transmitter.

To fight this parasite of power, I learned how to build a test set to measure standing waves. It consisted of a flashlight bulb, a loop of wire, a tuning *condenser,* and a broom stick. The idea, then, was to hold this loop, via the broomstick, somewhere near the transmission line going up the roof to the antenna. If the bulb lit, power was being wasted and adjustments had to be made.

It took some acrobatics to get close to the transmission line, which went right to the top of the roof and beyond. But I had no fear of heights at that age and soon found myself careening about at dizzy heights above the stately embassy building.

I was making progress on this scientific venture until, one day, I became too enthralled with my work and stepped back on a glass skylight. Crash, clang, clatter, and I probably looked like some part of a cop-and-robber slow motion film as I smashed through the skylight to the attic floor below. When the dust settled and the last clink of glass had sounded, I found myself flat on my back on the floor below. No broken bones – just a burning desire to cover up my *faux pas* from my mother and father.

In late 1939, our family left dear old Ireland for home, and FDR appointed my father to be our U.S. ambassador to Belgium, a most critical position given the amount of turmoil at the time. World War II was underway, Hitler was storming through Europe with a vengeance, while the U.S. was making at least some attempt to stay out of the conflict. The families of U.S. diplomats were not permitted to travel to Europe because of the risk, so my mother, my sister, and I stayed home in Milwaukee.

But my father was not to be in Belgium for long. The Third Reich was on a ruthless and relentless march to conquer that small, defenseless country, and nothing could have stopped them but an act of God. In the spring of 1940, Hitler's army had reached the outskirts of Brussels, and my father became the principal communicator to Roosevelt and our other ambassadors in Europe, reporting on every move of the attack. A press release from the State Department dated May 10, 1940, said, in part:

Drawing of Mike falling through embassy ceiling, Discovery World, Milwaukee

Ambassador John Cudahy informed Ambassador William C. Bullitt by telephone at 11:15 this morning (Belgian time) that the German forces had already overrun the whole of Luxembourg and all of Belgium. There was heavy fighting in the Ardennes. Ambassador Cudahy added that he had almost been knocked down by the force of a bomb which fell three hundred feet from the American Embassy and that one of his ears had been deafened by it and was still deaf. A number of windows in the Embassy had been shattered.

Later in May, my father met with Joachim Von Ribbentrop, Minister for Foreign Affairs and Doctor Joseph Goebbels, chief of the Propaganda Ministry in Germany. At that meeting, Father urged Von Ribbentrop to arrange an interview directly with Hitler so that he could address the concerns of the American people – that Germany might extend its desire for conquest to the United States. My father, however, was denied and told that such an interview might be viewed as a bid for peace by Hitler.

John Cudahy headed for America with deep emotional scars for King Leopold and the people of Belgium, whom he had grown to know and love in the short time he was there.

Back in the United States with his ambassadorial post gone, John Cudahy became a reporter for *Life* magazine. In the spring of 1941, his friend Clare Boothe Luce asked him to return to war torn Europe and report on the conditions over there and particularly on the status of the Belgian people. He was asked to return to Germany as well because it was felt then that an interview *could* be arranged directly with Hitler.

The interview, Hitler's only such meeting with any reporter in 1941 and beyond as far as I know, took place at Hitler's private hideaway in the Bavarian Alps called "The Eagle's Nest" near Berchtesgaden. It is reproduced in an appendix of this book (*see page 213*).

Fifty-five years after all of this, in 1995, I revisited Ireland, and the ambassador's home in Dublin. This time I arrived with the governor of Wisconsin, Tommy Thompson, a dear friend. The social secretary of the mansion was serving us tea in the drawing room on that occasion and decided to tell us a little story:

> You know, gentlemen, there was a very small lad years ago, the ambassador's son at the time I believe he was, who fell through a roof skylight of this lovely house, and tragically died! We know little else about the tale, but it was a very sad time indeed."

The governor, knowing the true story, and with a particular dry Irish wit himself, turned to the lady and said, "Madam, ghosts are with us today. The lad who fell from the roof those many years ago is still living and sits beside you as we speak!"

"Faith and b'gorey, do you mean to tell me that your Mr. Cudahy here was the boy who fell through the roof when his father was the ambassador in 1937?"

"He was, indeed," the governor said, "and that would tell you, you should be more careful with your stories about the olden days."

CHAPTER 3

WAR, LOVE, AND
A LIFELONG MENTOR

I perhaps, should have gone on to a formal education, as did most of my counterparts in the scientific world. But the war came and I was drafted into the Army Air Corps.

The adventure, along with the untimely death of my father, made this the most significant time in my life. First, in the eyes of any reasonable observer, I was a spoiled brat, and the service was about to knock this unattractive characteristic out of me within a matter of months.

Off I went to basic training in Lincoln, Nebraska, and I got there on a stripped-down, non-air-conditioned troop train in the middle of a very hot, midwestern summer. We were all in our late teens or early twenties, and most of the group were trying to show how *macho* they were by using every dirty and swear word known to man.

At camp, the first thing I found out was that I couldn't sleep in at my leisure like my dear mother always let me do.

"Fall in, you worthless bastards! Get your sorry asses out there and line up," the sergeant yelled. "Cudahy, one more problem outa you and you're on KP for a week." Unfortunately, I decided to call the sergeant's bluff and found out the hard way that he bore little resemblance to my sweet and kindly mother.

KP was awful. I was on pots and pans one time for a week after I decided to smart off. Yet every ugly thing that happened to me in those days ended up helping me for the rest of my life. I learned that the world was not always at my beck and call – that there were others out there who might have some say in how things might go and that when I was told to do something it meant "do it" – or else!

Basic training was three months long. Half way through, I got a frantic phone call from my mother that there had been a terrible accident involving my father. She couldn't seem to tell me if he was still living or not, only that he had fallen off his horse on our farm back in Wisconsin.

Almost immediately I was given a pass to return home on emergency leave and found, indeed, that he had died instantly, having been thrown by a young colt he was trying to train. He was riding without a hard hat and was thrown a considerable distance. The front page of the *Milwaukee Journal* covered the story with his portrait. He had been one of Milwaukee's better known citizens.

Back at basic training with my father gone from this earth, I felt a new and strange feeling that I was no longer a kid and that I had work to do in this world. More to the point, I wondered how I was going to survive in the miserable environment of the U.S. Army Air Corps.

As basic training ended – the most painful phase of my military career – we were all given at least one opportunity to express our desires as to what part of this great and wonderful organization we wished to serve in. Suddenly I spotted the words *RADIO AND COMMUNICATIONS TRAINING, Madison, Wisconsin* on one of the GI posters.

If only I could be so lucky. Madison, only eighty miles from home and training in the very subject I knew and loved more than anything else on earth. And, as luck would have it, I got what I prayed for, Truax Field Radio School in Madison.

I breezed through the courses with considerable ease. I had been studying all those very things for the last eight years of my young life and graduated with a 98 percent final mark, the top of my class. Then they asked if I had any desire to teach one of the classes and, of course, I jumped at the opportunity.

Later on, I was transferred to Chanute Field in Rantoul, Illinois where I attended RADAR school and essentially did the same thing as I had done in Madison – got a high grade and ended up teaching. Realistically, I suspect, the service was somewhat desperate for teachers at that time. I remember studying sometimes half the night so as not to make a fool of myself in class the next day. Although I knew the basics, this was RADAR, and there was one hell of a lot I didn't know.

At Chanute Field, I found myself continuously strapped for money. The pay for a PFC at that time was below slave wage, and I found myself inventing ways to make an extra buck, one way or another. I had a car, albeit a wartime clunker, but this was more than most GIs had. And I found the car was very useful in making money by various legitimate, and sometimes illegitimate, ways.

As an example of the latter, there were GIs on the base who could not get passes for one reason or another and were usually wanting to see their various girlfriends in town, sometimes desperately.

In a true entrepreneurial spirit, I discovered that I could smuggle these ladies through the security gates by locking them in the trunk of my wartime car, and, of course, I charged dearly for the service. Unfortunately, the practice usually kept me up until the wee hours since the young ladies would have to get back off base, and my service was the only way out. I never resorted to gouging, I hasten to add, just expensive fees.

I also ran a *bus* service to Milwaukee with my vehicle, carefully charging a premium price over train service since it was far more convenient for some. The car was always packed every weekend and, as a result, I always had money jingling in my pockets, which created another problem when the word got around.

All of this teaching activity managed to keep me from overseas duty, for which I was not sorry.

When the war was over, I returned to Milwaukee to ponder my options on the rest of my life. College occurred to me, but there was always something to do first. So, the sketchy education I got in the service, as well as my intense self-studying habits related to this radio hobby, made up my entire schooling.

I knew basic "electronics," as it was later called. I knew Ohm's law and Kirchhoff's law and how to design and build receivers and transmitters and how not to get electrocuted, although I came close a few times. And sometimes I found it a distinct advantage not to have had a formal education, which I will explain later.

Then I fell in love. Madly, hopelessly, passionately, insanely in love. We had to get married, immediately. Not because of a pregnancy. No, simply because we had to. Somehow if we didn't, we were sure that the moment would slip away.

I was 23. Mary Lee was an incredibly bright, fresh and pretty young girl of 22. We proved later that we were way too young for marriage.

My father had died some two years before and, although my mother begged us to wait a little while, we just went ahead and tied the knot anyway, to hell with Mother. Mary Lee's parents were shocked with our haste, but tolerated it by saying, "They'll live and learn."

I had three more wives after that, and I'm not particularly proud of it. In fact, it seems to demonstrate that I have been careless in my judgment of women, or intolerant, or immature, or impossible to live with, none of which appears to be a very nice quality.

It all was perfectly logical each time I stepped up to the altar – at least in my mind.

Marriage seemed such a dichotomy in my mind. Performed in the church and dissolved in a court of law, it required no preparation other than a rehearsal dinner. It required no training, no education, no psychological readiness. In elegant simplicity, it only asked for the truth between the parties when they pledge each other's love. It asked for this through sickness and health, for better and for worse, forever and ever.

Within a year or so of this first marriage, I started to feel the "live and learn" pain as if being a prisoner of war. Where had my freedom gone? Why couldn't I live the carefree life I had known before and my buddies continued to enjoy?

And then the baby came, my daughter, Susie. The shackles tightened even tighter. In utter disillusion, I ran for the escape hatch – divorce – and another single parent home was born.

The year was 1950, and I left good old Milwaukee again, this time for New York City. I ran off with another man's wife on my arm and decided that I was ready to expand my horizons. The lady I brought with me didn't move in with me, but rather with a bunch of girls I knew down on Bank Street in Manhattan, all of whom were cute and having lots of fun in the big city. One gal, Joyce, disappeared regularly at night in a big, black limousine, and I found out later it was to be with Senator Jack Kennedy. The others had somewhat less exciting affairs.

I soon tired of my cute little friend (or more likely she tired of me), and she went back to her husband in Minnesota to live happily ever after. I got a job at NBC television, a medium that was just starting, at Radio City. I worked as an audio man and later as an audio director. The pay was poor, but the script girls were beautiful, and I fell in love regularly every week. I lived up on the West Side of Central Park in Manhattan with a guy by the name of Dick Chenery, another Irishman. New York was nothing but fun in those days. We used to have *open house* parties (booze but almost no drugs) that were truly open – that is, we'd leave the door open, and those who weren't invited would walk in with those that were.

Living in New York City and working for NBC as an audio man gave me some free time to do something useful with the electronic skills I had accumulated through the previous years. The magnetic tape recorder had just been developed at the University of Illinois, an outgrowth of work done in Germany during the war on wire recording. Soon Magnachord of Chicago, Presto of Paramus, New Jersey, and Ampex of Redwood City, California were all in a race to capture a piece of a new technology market, that turned out to be truly explosive, but was later lost to the Japanese.

I had a wild and crazy friend by the name of Stuart Auer who was with the Muzak Company, originators of background music for stores, restaurants, and factories. One day he said to me "Why not put Muzak on airplanes? That way people will be more calm when flights are late or otherwise screwed up."

"Great idea," I told him, without seriously thinking about it. I said, "I'm just the guy who could do the job and Presto Recording, right there in Paramus, New Jersey, is one of the companies starting to build a new machine called a magnetic tape recorder. Maybe we could get one to work on a plane."

Soon I found myself going to Paramus every week to persuade the Presto Company to build a tape player specifically designed for use in an airplane. In fact, I told them I would help design and build the first one and even figure out a way to sell the device to the airlines.

Presto went along with the idea and even put some engineers on the job to help me. We built the one for National Airlines in Miami, and soon they were equipping much of their fleet with this new and wonderful way to soothe passengers.

In the meantime, one of Presto's reps, a fellow by the name of Warren Cozzens, got wind of the idea and called me.

"Your name Cudahy? You the guy working on a magnetic tape player for airplanes?"

"Why, yes, we have one on a National plane as we speak. What do you have in mind?"

"I'm out here in Chicago, and all the big railroads are head-quartered here. Having music on railroad cars, like diners, is even better than music on planes."

"Could be," I said, "but what's your connection?"

"I'm the Presto rep out here, and I'm always lookin' for a sale, that's what."

"Sounds like an interesting thought, Cozzens, why don't I come out there sometime and see you?"

"Take the Grand Central overnight train tonight. I'll pick you up at Union Station in the morning." Click!

I arrived in Chicago at 8:30 AM and there was Warren B. Cozzens, a guy you couldn't miss in a crowd. He was six-foot-four, blond hair and looking like he owned Union Station. "Cudahy? I thought so. You described yourself well. My car is illegally parked so we better get the hell out of here before I get a ticket."

Warren B. Cozzens

"Let's go," I said, and suddenly I felt like I'd known this guy all my life. We went directly to his house in Evanston, and he introduced me to his wife, Barbara, who asked, "Did Warren tell you you're staying at the house tonight?"

"No, but that's fine with me."

In the weeks that followed, Warren and I became close friends, and soon he persuaded me to give up New York, my job at NBC, and move to Chicago to join his rep company. Why not? I had no ties in New York. No girlfriend that I couldn't live without, and I was a bit fed up with the big city. I looked at the opportunities that I might have in the good old Midwest, and my gut said this guy Cozzens was pretty much OK.

My relationship with Warren brought about the second major change in my outlook on life (the first being basic training in the Air Corps). Warren was the guy who taught me the fundamentals of business, and how to live in the world. When he died in 1998, I gave a eulogy in which I said:

> I'll start by telling you he was the best friend I have ever had. He was the "mentor of all mentors" as I told the newspapers, and today I'd like to tell you why.

Cozzens mentoring Cudahy

Then I told about his original phone call and my decision to move to Evanston.

Continuing with my eulogy, I said:

> He and his wife Barbara used to say I was "the man who came to dinner" because I moved right into their spare bedroom and wouldn't find a place of my own. Finally, they had another baby and I had to move. That might have been why they had the baby, come to think of it.
>
> And Warren filled me with his business one-liners like "get the order, and then decide if you want it" – or "an order isn't an order until you've shipped it and you've been paid" – or "never argue about money – it'll take care of itself if you do your job right."

Warren was my best man in two of my many weddings, and I was the best man in his when he got married to Jean after Barbara died.

The world is a much, much better place because of Warren Cozzens. He has mentored and improved the lives of hundreds of people, and he took this mixed-up kid so many years ago and made me what I am today – a successful graduate student of the Warren B. Cozzens school. Instead of MD or PhD, perhaps I should have a "WBC" after my name.

May God bless you, Warren B. Cozzens.

Warren had four children. Chris, a PhD in English, Jeff, a neurosurgeon, Todd, and Michael, my namesake. Mike runs a resort in northern Wisconsin, and Todd is the subject of a later chapter.

In 1959, we moved the Cozzens and Cudahy rep firm from Evanston to a shopping center in Skokie, Illinois, and for the first time, we had a few extra square feet compared to our cramped quarters in Evanston. Warren was trying to quit smoking, and I was thinking I should be married again. We represented some great electronic companies back then, like the John Fluke Manufacturing Company of Seattle, Washington, and I'd say we were doing well, all things considered. We went to a training seminar given by the Fluke Company that year at a resort at Hood Canal.

The boss, "Big John" Fluke, was some sort of character, and I loved everything he did and said. Six-foot-four, balding and a bit heavy, he was loud and funny and profane and frightening to most of the gang. He was a graduate of MIT and had been a Navy commander in the Pacific in World War II. When the Korean War came along, he used to say, "Why don't the stupid bastards just nuke the sons of bitches?" Not that that was so smart – he just had a very direct way about him and an IQ somewhere between a thousand and infinity.

For some reason, I didn't share the fear most had of Big John and usually tried (but always failed) to outsmart and out dirty-talk him at these meetings. When I visited his plant, which was two or three times a year, Big John would take me through the place and show me the latest machines he had installed for improving his production.

"What's that monster, John?" I asked during one visit, gazing up in awe at a fifteen foot high, ten ton machine that had just been installed on the plant floor.

"Transformer winding machine, stupid! Just got it. Going to make one hell of a difference around here with our quality."

"You mean wind your own transformers? Why, there must be fifty local companies around here making stuff like that what with Boeing Aircraft being right around the corner," I said.

"Yeah, Cudahy, you're right," he said, "and they all make crap! 50 percent of our failures in the field are the lousy transformers made by those turkeys, and now I can kiss all the stupid bastards goodbye!"

How fantastic, I thought. This guy just doesn't screw around. He does what needs to be done and gets on with it. How many people study and study problems like that and then don't make a clear-cut decision. Big John was my hero, and if I ever had a manufacturing company, I'd run it just like John's.

Warren and I were on the same flight together back to Chicago from the Fluke meeting, and the conversation got into what we really wanted to do for the rest of our lives. We had had it up to here with the Fluke meeting although we loved the company and the old man. It was just *enough is enough* talk from three days of high intensity.

Warren said, "You know, Cudahy, you and I are like the play *The Death of a Salesman*. Do you remember that show?"

"Yeah, that's the one where the poor old peddler has lost his girlfriend and his best account, and he's having his tenth drink and feeling like killing himself, right?"

"Uh yeah, sort of, that's the one, and someday we're going to be just like that guy with nothing to show for all our work but a few old typewriters and some used furniture."

"What's got into you, Cozzens? Why do you act like you lost your last friend? Why do you think we're going to end up like that?"

"Ah, I guess I'm just tired," said old Warren, but he kept that sort of talk up for more than half the trip. Finally I said, "Maybe we ought to be manufacturers like Fluke, Warren. That way we'd build some assets beyond typewriters and used furniture. We're just as smart as Big John, and there are all sorts of things we could manufacture – things we know a hell of a lot about, agreed?"

"Yeah, maybe you're right, Cudahy. Maybe we shouldn't just sit around and mope. Maybe we should start a manufacturing company and sort of ease ourselves out of this rep business."

"Well, I'm not the guy who's moping, Warren, but I think it's a great idea. Seriously, I think we should do it. I could start it up in Milwaukee, and you could keep the rep outfit together in Chicago so we can keep on eating until the thing starts to make money, OK?"

"Milwaukee? OK, I guess that's where you really want to live anyway, Mike, so why not? Let's give it a shot. What should we manufacture?"

"Details, Warren, details. We'll find something to build – I'll give it some thought. Maybe we could build RF signal generators. You're always griping about how nobody makes a good generator for the 400 megacycle band. Sure, we could do that. Probably make a fortune at it."

"Cudahy, you have the damnedest way of making everything seem so easy. Building one of those machines is not as simple as it looks, you know. And who's going to engineer it? Certainly not you."

"Ah, Cozzens, have another drink. I'll flag down the stewardess."

The T-BASS, Cudahy's electronic bass fiddle

Chapter 4

MARQUETTE IS BORN

And so at last the idea of Marquette was born. No business plan. No product plan.

Almost no money. No engineers. Just an idea that if we didn't do something, we'd end up like the guy in *Death of a Salesman*.

I'm not poo-pooing business *plans*. Obviously, it's nice to have such a thing when a new company is started, but it's people that make a company happen, not a plan. I have seen so many start up companies create these plans and then, somehow, expect things to just magically happen that way.

We made a plan to go into the manufacturing business, but for almost a year we fiddled around with various possibilities of things to build without much success. The radio frequency (RF) signal generator we had considered was, just as Warren Cozzens predicted, harder to design than we had expected, and I wondered if it was worth it.

I have an old friend by the name of Ray Brown, who is one of the world's most famous jazz bass players. One day, he and I cooked up the idea of building an electronic bass fiddle you could play with your left foot while also playing the piano. Some might say that wouldn't be a very nice thing to build since it could put a lot of bass players out of work, but as it turned out, they weren't in any danger.

I found out that another friend and great jazz musician, Claude Jones, was playing some sort of jury-rigged electronic bass gadget down in Chicago, so one night I went down there to listen, and when I did, I told Claude I thought his gadget sounded like hell.

"Ray Brown and I are building something much better," I said, and his only comment was, "Well, why don't I try yours out if you think this rig sounds so bad?"

Ray and I plunged ahead until we thought we had something that sounded pretty nearly like the real thing – Ray in person, that is. Claude bought the very first unit for something like $300, and I proceeded to *manufacture* nine more and actually sold a few. Problem was that the cost-of-goods was about $400, and that didn't include labor, which I threw in without charge. It didn't seem that I could charge a penny more from what musicians were telling me.

So the T-BASS, as it was called, went by the wayside, except for Claude, who put three kids through college playing the unit he bought. Claude, by the way, was the only musician who ever paid for his T-BASS. Musicians are not always known for their fiscal responsibility, I found out.

In any case, I wasn't discouraged. I was convinced I'd find another product that would someday make our manufacturing venture a success.

Back at the rep company in Chicago, Warren and I were walking out the door for lunch one day when, as usual, the phone rang.

"Joan'll get it, Warren. Come on, I'm hungry," I said.

"No, Mike, we'd better wait. It might be a customer just itching to buy something."

"Oh, screw it, Warnie. Come on, you always say that."

"Nope, Mike," said old Warren. "I've got a feeling."

Then Joan called to us. "It's some Doctor Winnemark who wants info on a large screen memory scope," and Warren went to pick up the call.

"No, doctor, we no longer represent Hughes Aircraft," Warren was saying, and he gave the good doctor on the other end of the line the name of the new agent representing Hughes.

Then Warnie's inquisitive brain kicked in as he asked, "Say, just what are you trying to do with a Hughes memory type oscilloscope if I may ask? Record electrocardiographs? Uh huh. Well, you really don't need a memory scope for that. If you're going to put it on film anyway, the film will be your memory."

Electrocardiograms are recordings of minute electrical signals from the heart. These recordings are very useful in determining the condition of this amazing pump inside the human chest.

Until we entered the picture, electrocardiograms, EKGs (or sometimes called ECGs) were recorded on a small machine that was wheeled around the hospital to the patient's bedside. Once there, electrodes (small metal discs) were placed on the chest with *electrode paste* and wires connected the signals to an amplifier and a paper chart or *strip* recorder on the machine.

After getting the patient wired up, the *strip* recorder would produce some eight or even ten feet of three inch wide paper in a roll. The technician would then stuff this record in her pocket and return to the heart station.

At the heart station, other technicians would cut and paste these recordings on letter sized pieces of cardboard so that cardiologists could make their *interpretations* of the EKGs.

With the scheme these doctors were proposing, a new kind of machine would be wheeled to the bedside that would acquire the patient's heart signals in the same way but then would transmit them by modem over the telephone line to the heart station. Once there, a central processor would automatically prepare the record for reading by the doctor, saving vast amounts of time and labor.

Warren and I knew nothing about electrocardiography. But the doctor must have been impressed with Warren, because the next thing I knew he was making a luncheon date with this guy, apparently to talk some more about the idea.

A few more meetings and we had an order to build the world's first central electrocardiograph system. Dr. Winnemark and his associate, Dr. Zitnik, had already tried Cambridge Instruments and Hewlett-Packard, the major companies in that kind of business, but they both said the idea was totally impractical.

Looking back, if good old Warren hadn't have been so inquisitive that fateful day, I think it's fair to say that Marquette Electronics would not have been in the medical electronics business today. Nor, perhaps, in business at all. Fortunately, 1964 was the beginning of the golden era of medical electronics, and medicine was in need of partners to help with the discoveries and procedures that were emerging so rapidly. And there was no Food and Drug Administration control over *devices*, no Underwriters Laboratories, no requirement for clinical trials – nothing but just "try it and if it works, go for it!"

We were willing to try. All we needed now was a place to design and make this strange machine.

The first home for our new *manufacturing outfit* was truly humble. So many starting companies go directly to venture capitalists before even opening their doors, and thus have the money to rent or buy a nice place and acquire fancy furniture. We decided that things like that had little, if anything, to do with beginning success.

We rented a very small building on Abert Place, a sort of alley a few blocks south of Capitol Drive in Milwaukee. Its total of

Marquette Electronics' first building on Abert Place, Milwaukee

1,600 square feet included the furnace room. There were two wood and glass partitioned offices at one end, and that left an open space of about 1,200 feet for *engineering and assembly*, if you could call it that.

Furniture? Some came with the place. Old wooden desks, a few tables and one or two rusty filing cabinets. I brought some old junk from my basement at home to fill in. Linoleum floors, in bad shape, throughout. It was air-conditioned, believe it or not, and that's why the rent was so high – $550 per month. The electrical circuits were very limited, however, and when someone used the copy machine (a 3M Thermo-Fax) we had to unplug the soldering irons or the fuse would blow.

When we got a little further along, I invited our two doctor friends, Zitnik and Winnemark, up from Northwestern University Medical School and was very nervous about what they'd think when they saw the place. They were very forgiving, however, and much more interested in the machine we were building for them than our humble quarters. Later on, when we acquired a few other

The original Marquette Electronics gang (left to right), Gil Johnson, Chuck Pearson, Mike Cudahy, Joan O'Hara, Warren Cozzens, Ron Shively and Dan Phillips

customers, they liked to kid us about the place although I doubt if we ever lost an order because of it.

I hired a technician by the name of Dan Phillips, who had been working for the Delco Division of General Motors in Milwaukee. He was fed up with all the bureaucracy over there and wanted a change – quite a risk to come with us since he had four hungry kids and a wife at home.

Dan was one of those guys who could always *get it for you whole-sale* no matter what it was. So when we decided to paint our name on the outside of the building, Dan, naturally, had a friend who could do the job much cheaper than a *regular* sign painter. I hired him on Dan's say-so, and the next week this guy, Elroy, shows up with paint, brushes, ladder and good looking girlfriend – to hold the ladder steady while he painted.

We had decided to name the company Marquette Electronics because Warren and I thought that would give the place some sort of *intellectual class*. Besides, I knew the president of Marquette University, Father Kelly, so I figured I could get his permission.

Elroy, the sign painter, proceeded to carefully lay out "Marquette Electronics, Inc." on the very top cement blocks of the building. Dan and I didn't pay much attention to him except to glance at Elroy's girlfriend from time to time and make lascivious remarks.

The next day, the mailman made his usual morning delivery and said, "Say, I thought you fellas named this outfit after our local university."

"We did, Bert, why do you ask?"

"Well, you sure ain't spelling it like the university," our friendly mailman remarked.

Dan and I immediately rushed outside, and there in big, bold print was MARQUEUTTE ELECTRONICS.

"Good God, Dan," I said. "How stupid can the guy be? He's got an extra U in Marquette. What do we do now?"

"Don't worry, Mike," said Dan. "I'll get the guy back tomorrow and have him fix it."

"It's not going to be easy matching that bluish paint on the cement block, Dan. All I have to say is 'good luck.' Tell the dumb bastard it better look perfect or he's not going to get his money!"

"Already paid him, Mike, but I'm sure he'll make it right."

Elroy was back the next day and tried and tried to match the blue cement block, but the more he patched, the worse it looked. Finally, he had to repaint the entire side of the building. The girlfriend called him an idiot, and we felt so sorry for him that we scrounged up an extra $50 to pay for all the extra paint.

Soon we sat down to see if we couldn't figure out how to construct this central electrocardiographic system the doctors wanted. They said it would revolutionize the world, and we said sure, but first we had to see if we could build it and make it work. Transistors were in, vacuum tubes were out, so we said, "Let's be really modern and make it all with transistors." Problem was, the only kind you could get in those days were germanium transistors, and they had a tendency to drift all over the map with the slightest temperature change. Silicon just wasn't ready and available in 1964. Oh well, we said, the system would be going into an air-conditioned hospital, and maybe things wouldn't be too bad. Dan and I knew a few things about transistor circuit design, but only a few.

Data acquisition cart? Something to wheel around the hospital to pick up the patient's EKG? Piece of cake we thought. Nothing to it. The first acquisition trolley could be made of some standard bits and pieces we could purchase from the Sanborn

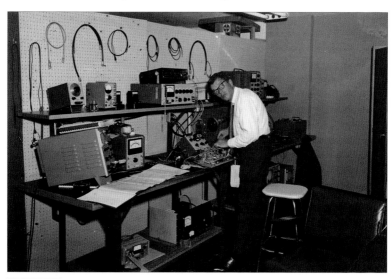

Mike tinkers.

Company, a new division of Hewlett-Packard. We'd add a *modem* and for that we prevailed on good old Ma Bell and the Bell Labs because Warren had a friend there by the name of Joe Ennenbach. Our central system, then, would need a microfilm camera to store wave forms appearing on the picture tube. We found that from the 3M Company.

And it was thus that four unlikely partners – Hewlett-Packard, Bell Labs, 3M, and Marquette – came together to make the project happen – one way or the other. Painfully and slowly, we worked out the circuits for the central control system, redesigning often as we went along when various test circuits worked or didn't work. Trial and error in the true sense of the word. My qualifications as an engineer were only from being a ham radio operator and my schooling and teaching experiences in the Air Corps. Dan knew less than I did. He had been a technician at Delco, but only that.

Yet we inched along each day, calling engineering friends from time to time for what free advice we could get. Both of us began the assembly of this *monster* at the same time as the circuit designs began to take shape.

Assembly proved more difficult than the design, believe it or not. Part of the problem was that I was on the road a lot scrounging up parts. Partner Warren Cozzens remained in Chicago running our rep outfit, and Dan had a bit of a drinking problem. It seems by the time he could figure out where he left off the day before on the complex wiring job, it would be time to go out and have a beer and some lunch. Then it took most of the afternoon to find his way back into the maze of wires.

One day it dawned on me that we were going to lose our shirts on this deal. We had much more labor in it than I ever imagined, and the parts we had to buy were much more expensive than anticipated. We had originally settled with Northwestern for a total

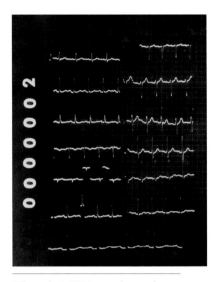

The author's EKG, second ever taken on Marquette equipment

price of $12,000, and I now figured we needed $30,000 to break even. On bended knee, I went to see the administrator of Northwestern and told him my problem. Much to my surprise, he agreed to "double the price – and not one penny more."

Fantastic, I thought, now I'll only lose $6,000. But soon Northwestern began to badger us about a delivery date, and I started to realize I'd better get another tech into the outfit to help poor old Dan if we ever expected to make something that would work and we could take to the hospital. The other suppliers – HP, 3M, and the Bell Labs – were ready to deliver their parts, mostly because they really didn't have to develop anything. So we were in the hot seat, and I hired employee number three, Ron Shively.

Delivery was made December 1, 1964. Ron, Dan, and I rented a truck and drove it to Chicago with a mixture of joy and panic in our hearts. The good doctors were, as expected, waiting anxiously at the door. A few hours of grunts and groans and we were there. Now to hook it up and get our money.

Little I knew of the trials and tribulations that lay ahead. Little did I know how long it would be before we would get our money, much less avoid taking the whole thing back and being sued for fraud. The germanium transistors drifted, in spite of the air conditioning. The camera interface failed, and 3M said it was

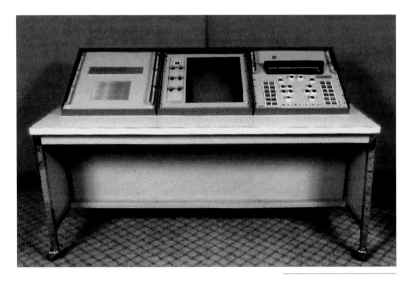

First Central EKG System

all our fault. The modems that Bell Labs made didn't work worth a hoot, and the Hewlett-Packard trolley was heavy and very temperamental. We got all the blame for all the problems; at least, that's how it seemed. The doctors were patient but said we'd best get our act together or take it away.

And, little by painful little, we fixed one problem after the other until we managed to talk them out of at least part of the money. But we had to promise to keep working on it until it was perfect.

After that, things got so bad we decided to bring the whole bloody machine back to Milwaukee. When we got it there, we swore we would get every bug out of it before we'd return it, no matter how long it took.

It took a long, long time, and the phone rang off the hook every day. The doctors came to see us once a week. Life was hell, and I wondered why I'd ever left the rep business and why

Warren had ever answered that fateful phone call from Dr. Winnemark.

But one day we decided, although it wasn't perfect, the world's first central electrocardiographic system had to go back to its rightful home. Soon after that, the heart station at Northwestern started to process electrocardiograms coming in by telephone from the surrounding hospitals of Passavant and Wesley. Even a few staff members were impressed.

By the fall of 1965 the system was working even better, but if it had been an airplane, I can tell you, I wouldn't have flown in it. We had our good days and we had our bad. Hans Wessel, the doctor in charge, would call only when there was a crisis, at least in his opinion, and so we dreaded phone calls from him like the plague.

One day he called, not to report a breakdown, but to report something very strange going on at the center. He said that people from the 3M Company had been there, not to check on their microfilm camera but, instead, to look over our circuit boards with an eye to seeing how they worked – in other words, spying. Now, I always thought Hans was a bit paranoid, so I really didn't get too excited. It seemed very far-fetched to me that a big company like 3M would stoop to such a thing, and for what? To make a similar system? To go into competition with our system since they made the camera? Crazy, I thought, it must be Hans' paranoia.

But as time went on, more and more reports started to trickle back from Chicago that various, not-too-subtle people from 3M were snooping and even taking parts out of our machine to have a good look. I had become good friends with a very cute technician down there, and she was giving me regular reports on this espionage activity, and she didn't suffer from the same suspicious mind-set as Hans, believe me.

Many months later, we found out what was really going on with 3M. We had decided to go for broke by exhibiting our equipment at the American Heart Association meeting in Washington DC that fall, and we got a list of other exhibitors as a part of registering. Low and behold, there it was in print – 3M was going to be there, too, with a new central electrocardiographic recording system.

"Bastards!" I screamed to all three of our employees. "The dirty bastards have copied our system and are going into competition with us."

And they did. Full bore. Their central system. That's what they had been up to at Northwestern all along. They had blatantly gone into our system and copied our circuits and apparently decided it might be a good way to sell cameras. Who needs Marquette Electronics, they apparently thought.

Soon the American Heart meeting was upon us, and we had built serial number two in our tiny shop with many improvements over the unit at Northwestern. We thought it was pretty slick. We even managed to get rid of the germanium transistors and replace them with silicon.

We packed up the system and rented another truck, this time to get our baby to the airport where Northwest Airlines would fly it to Washington. Dave Ivers, our new service guy, Ron Shively and I caught the earlier flight so we could be right there when our precious cargo arrived. Another truck rental at the airport in Washington, and we were at the convention center a full fifteen hours before show time.

Luck was with us that day. We unloaded and set up the system and by nine o'clock that evening, we had everything working perfectly. Not more than a hundred feet from our little, humble spot was a huge and elaborate booth, complete with a tall, gold carpeted pedestal at the center. The brand name 3M

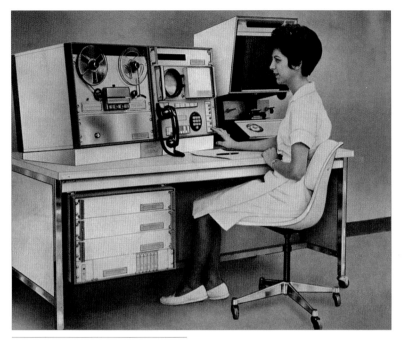

Marquette's first Production EKG System

shown in bright letters on an elegant new machine – the 3M Central ECG Recording System, model 5000.

"OK," I said, "tomorrow, we'll take a look and see what they have. Let's go and have a drink. There are too many of their creepy guys standing around right now." And we took off with a touch of uneasiness but still pretty cocky about how well our own equipment was working.

"Let's come back later tonight," said Shively. "They'll all be in bed, and we can have a good look."

"Yeah," said Ivers, and we agreed we'd be back at eleven. If the show officials gave us a bad time, we'd tell them we were having trouble with our equipment and had to get it fixed before show time in the morning.

When we returned to the convention hall, perhaps a bit in the bag, we found to our utter amazement that the 3M Company had surrounded their entire booth with a tall, dark blue curtain, suspended by an elaborate set of piping. A Pinkerton guard paced up and down on the outside and lots of cussing and cigar smoke came from within. "Hand me a screwdriver, asshole," someone bellowed, and then, "Fire it up again, Walter. We're running out of time," came another voice, and then, "Oh shit!" from a third.

What ever possessed me to do what I did next I do not know. Perhaps the booze. But in a moment of total defiance, I pulled back a section of the curtain and said, "Need some help in here?"

A big guy came flying up to me and said, "Who the hell are you? Get out!" And now the Pinkerton man was approaching and looked ready to do something nasty. "This booth is under security," he announced, and we decided we'd better get moving.

The show opened promptly the next morning, and our equipment performed flawlessly. 3M had removed their blue curtain and proudly greeted their anxious customers as well. The night's labor had apparently paid off, and they seemed to have their machine working. A very pretty model had the seat of honor up on the pedestal where she confidently pushed the master control buttons and showed a raft of curious customers the new miracle 3M had built.

But soon one of our many cardiologist friends came over to our booth and gave us some wonderful news. They joyously announced to us that the 3M machine was not functioning at all – that the good-looking blond model was trying desperately to fake it. No microfilm pictures were emerging from their camera! When our friends asked 3M what was wrong, the salesman attending the demonstration said some minor adjustments were being made, and all would be well in a few minutes.

It never was. Adjustments were frantic, but apparently fruitless. Each night, when the exhibits closed, the big blue curtain would reappear, and the Pinkerton man would begin his guard duty. The cigar smokers would work through the night to straighten out the problems, but as near as we could determine, the great new machine never did work. We, of course, continued to demonstrate our well-functioning system, and I'd be the first to admit that we were a bit smug.

The doctors? Well, they developed a rousing support for us, the underdogs who had been attacked by big, bad 3M.

CHAPTER 5

INNOVATION, OUR LIFE'S BLOOD

I n 1967, I finally had the wisdom to realize that I was not
going to be able to do all the engineering for this company on
my own. As an almost-high-school-drop-out, ham-radio
operator, I could design electronic circuits, but they were always
seat-of-your-pants in performance and couldn't be trusted with
vital medical data.

With this in mind, I set out to hire a chief engineer, and
rather quickly came upon a guy from GM's Delco division in
Milwaukee by the name of Gil Johnson. He was fed up with
Delco's bureaucratic ways and wanted to spread his wings a bit,
or so he said.

Nowhere on God's green earth has there ever been such a
straight-as-an-arrow guy, and never has there been a better electrical engineer – at least for his time. He had graduated from the
University of Wisconsin in 1960 with honors and was a true engineer's engineer, if you know what I mean. There was only one way
to do a job, and that was the right way – by the books. Any sug-
gestion of a short cut to making a product work, and old Gil would
fly off the handle with a vengeance.

One day I came in from my travels and concluded that electrocardiographs were all wrong and that to make them right their
amplifiers would have to be DC coupled. This may not mean

Gil Johnson

anything to you, but it did to old Gil, and he promptly said it couldn't be done. No ifs, ands or buts, it could not be done, period.

Being a guy with a certain amount of persistence, I kept on nagging that we had to find a way to do this if we were going to make a true contribution in the world of electrocardiography, and one day Gil totally lost his cool. He went in his office and slammed the door with such force I thought the thing was going to split in two, and he stayed in there until after we all went home.

The next day we found the door still closed, locked as a matter of fact, and became somewhat worried about Gil's welfare, wondering if he'd been in there all night. I finally got up my nerve and gently knocked, asking if he was OK.

"Leave me alone," he said, and in a way that left us no choice but to do just that. One of the guys said Gil had sneaked into the lab for awhile just before noon, but had returned to his office before anyone had a chance to ask him what was up.

This sort of hide-and-go-seek conduct went on, more or less, for three days at which time Gil suddenly appeared at my office and asked if I'd like to see something in the lab.

"Sure, Gil, I'd like to see anything you're doing. I've been a little concerned about you, you know."

"Never mind that," he said gruffly. "Come in here and see your damn DC coupling if you want."

And there it was, bigger than life itself, the world's first DC coupled electrocardiograph, working perfectly before my very eyes. Gil stood there as proud as a new father. Shortly after that, we were able to produce Gil's invention, and our company became quite famous for this valuable, exclusive feature.

Later, when things became more tranquil between Gil and me, he said, "You know, you have a certain advantage over me without a formal education."

"Huh?" I said, wondering what kind of a bait that remark was, and before I could figure it out, Gil volunteered an explanation. He said that, in his view, a formal education inhibits an inventor because it tells him exactly what can be done and what can't. "If a formally educated guy thinks of something off the beaten track," he went on, "then his education will tell him that's not the way it's done."

I've thought about those words of Gil's many times since then, and the more I think of them, the more sense they make. Some well-known inventors and entrepreneurs have had very little education. Tom Edison for one, and right in my own hometown of Milwaukee, the Bradley brothers who started Allen-Bradley. I've thought that, perhaps, educators should realize the fact that there's this downside to a formal education and thus, have courses and lab classes in creativity. Perhaps some do.

The juices that seem to brew innovation were certainly present in the minds and the genes of our forefathers – Edison, with his light bulb; Ford and the automobile; Farnsworth with his picture tube for television; and Bell with the telephone. I think it might be fair to say America's wealth and world dominance in technology came from these men and many more who found, for some reason, that America provided a setting ideal for creativity.

Was it *freedom*? Was it getting away from the formal structure so prevalent in Europe at the time? Or was it that the time had come for these inventions to happen? In any case, it is my considered judgment that America has lost some of this spirit. And, although many wonderful innovations are still emerging from our society, I feel much of our best creativity is being absorbed with the quest for the *bottom line* and other shallow goals.

Without question, a continuous flow of innovation was a major factor in our growth from our humble beginning on Abert Place in Milwaukee to our worldwide success with gross sales of almost $600 million and employment of 3,300 people. Throughout the 34 years of our history, we had good times and bad, recessions, regulatory changes that sometimes hit pretty hard, but our innovative products always seemed to save the day. Competition was forever trying to catch up. As soon as they did, we'd have something new.

Innovation, and what makes people create truly new ideas has had me fascinated for years. I have been described by some as an innovator, but I'm not so sure I qualify. In the first place, we've had many an engineer working for us that innovated or invented new things, and entirely without my help. I was the root of *some* of our new ideas, but that's because I spent many a day with customers, finding out what they needed in the way of useful machines, so it seemed to many that I was a great source of creativity. In honesty, the real source of innovation was the customers we had and our close relationship with them.

A new idea often comes from the prompting of others, like customers. I used to come home from a long trip just chuck-full of ideas from doctors and nurses who I found had real needs. When I'd get home my problem was to convince our engineers that what was wanted could actually be made into a product. Not all our engineers were that way, but quite a few, in my experience, had a tendency to say "that'll never work" long before it had been fully analyzed.

Here are a few factors, elements I think make up the right (or wrong) conditions for encouraging innovation or invention.

A NEED

Sounds simple, but there have been many inventions laboriously created where there was little or no real need. Yet what was the need for the automobile before it was invented? Was the need to get around faster or better than with the horse and buggy? And what about television? Did we really need it? Need helps sometimes, but some inventions seem to just come out of the blue.

A CREATIVE MIND

This might be someone who thinks up an idea from scratch or finds a need and then creates a solution. Who has such a mind? I am still unable to judge. Is it someone who can think "off the beaten track" as Gil Johnson said, perhaps someone who is restless – unsatisfied with things as they are, or someone who feels there has to be a better way?

A PASSION

The person or the people who invent or innovate must be impassioned. This could be an internal thing or might simply be to show family and friends how smart they are. I never got into the *why* part with people. I was just interested in finding out how strong the passion really was. My experience tells me there are a lot of people out there who give up too easily. They'll go just so far, and then find excuses as to why it can't be done. Or, they'll listen to the naysayers who take joy in saying "I told you so" when it fails. The successful *creator* is someone who grits his teeth together when the going gets tough and says, "I will not fail. I'm going to do it, one way or the other."

FACILITIES

In 1982, we built a beautiful research center near our main plant in Milwaukee, creating an atmosphere (we hoped) that would inspire engineers to think *outside the box*. It worked – at least to some extent. It pulled talented engineers away from the humdrum of our everyday activities so that they could concentrate. This helped when the objective was more or less defined, but the basic need still had to come from the field – the potential user – the customer.

Main plant employees used to gather around and declare that people up at R&D did nothing but watch the squirrels (it was located in a beautiful wooded area in the northwestern corner of Milwaukee) so I gathered up some facts about the place. I found, to my delight, that some of the greatest innovations the company ever introduced came from our Research Center. I would hasten to add, however, that someone's old, dirty garage would have worked just as well if the right people, the tools and instruments were made available.

MONEY

How many times I have heard, "If we only had the money we could go ahead with this project." Surely, there comes a time when money is absolutely necessary to proceed. Test instruments, tooling, computers, special talented people – and all sorts of things require money – sometimes. But in the same breath, I'd say that the lack of money often is used as a prime excuse to declare, "We can't do it. No money." We all knew Tom Edison surrounded himself with wealthy people, like Henry Ford and Harvey Firestone, but the Wright brothers did their thing in a humble bike shop.

Here's a wonderful example of a guy who would not be swayed by anyone or anything in the pursuit of his innovation:

Jim Graham, a brilliant, but very reserved engineer, came to me one day and said he had an idea for a gas analyzer (to be used in operating rooms) that, he said, could be made smaller and cheaper than anything then existing. He asked if he could have a go at it, and I had a sense he was onto something. I said, "Get yourself a team together, Jim, and make it happen!"

Perhaps, then, too much time passed, the team grew larger and our then CEO came to me one day announcing that Jimmy's gadget wasn't working and, furthermore, would never work. The CEO said we were wasting a ton of money with seven engineers working at it. I listened carefully since this fellow was a PhD in physiology, an electrical engineer – and just might be right. I then invited poor old Jim out for lunch to probe his mind and see if he was losing it or his determination. He looked over his lunch at me with a sad eye that seemed to say "I will not fail if you'll trust me" and, before I knew it I said "just keep on going, Jim."

Six months later I was invited to watch the "SAM" module (as we called it) perform, and perform it did, with proud Jimmy and his team standing by.

A year and a half later we had sold over $6 million SAM modules and to some of the most prestigious hospitals in the world.

I referred to the *need* of an innovation or invention above. Is it really needed or is it just a figment of the inventor's imagination? Perhaps a customer's imagination. Customers and inventors alike occasionally find themselves so immersed in their idea that they lose sight of reality. I have actually designed (or had designed) an instrument by request of a doctor only to find that he really didn't need it after all. I have revisited doctors with some gadget in hand that they requested only to have them say, "I wanted that? I must have been off my trolley that day. No, we couldn't ever use such a device."

In the beginning, I was the guy who went out there and made friends with key people in the medical world to probe what was thought to be needed, especially by applying new technology. And it was I who screened the ideas I would glean from doctors, administrators, and nurses, grinding ideas up in my mind until a true product need would emerge, or at least I hoped so. Some of this was guesswork, I'll admit, but, somehow, I seemed to be more right than wrong – perhaps 51 to 49 percent.

As the company grew, however, I was faced with the task of finding others to do this work – engineers, salesmen, customer contact specialists, whomever we could find. And how to teach them this skill I had acquired through the years became a formidable task – how to get them to be trusted friends with the right doctors as I had done – how to get them to judge which doctors out there to listen to and which ones to ignore.

In the first place, one must, over a period of time, instill a feeling of trust with customers that must run deep, very deep. I had no difficulty in this regard because I was always painfully honest with them, and somehow they knew it. But I found this simple rule was difficult to teach. The trainee would always say, "Well, you mean reasonably honest, don't you? Surely, you can't tell them everything."

"Yes, you can," I'd say. "I mean everything! The raw truth – be it good or be it bad." I've been known to sit with my favorite salesman in front of a prized customer and say, "I hate to tell you this, Doctor, we tried to do what you wanted but we couldn't make it work," whereupon the salesman would collapse with embarrassment.

CHAPTER 6

TIMING IS EVERYTHING

How lucky we were to have fallen into a part of medicine that was still in a state more of witchcraft than of science, a part of medicine that was hungry for modern technology, with its potential to advance efficiency and standardization.

The electrocardiograph in 1964 was respected but ensconced in the special skills of a chosen few called "electrocardiographers." They read and interpreted the complex wave forms of the heart's electrical activity on graph paper, using calipers to make their measurements.

The art, or perhaps now the science, of electrocardiography is still read by a skilled electrocardiographer, but almost always assisted now by the modern computer. Often called an EKG, it is one of the most significant tests in medicine.

I say "still" because electrocardiography dates way back to the latter part of the nineteenth century and a scientist named Wilhelm Eindhoven of the Netherlands. Eindhoven is credited by most for the development of the first truly useful method for detecting the minute electrical signals of the heart that exist on the surface of the skin and recording them on film.

It seems that in the infinite and wonderful world of nature, the myocardium (heart) is driven by electrical impulses from what is called the "AV Node." From there, these tiny pulses, human and animal hearts of all kinds, precisely time the many activities of their chambers to pump blood through the lungs and body.

Marquette was at the doorstep of making the recording of these signals easier and better than ever before, although I'm not at all sure we recognized that fact at the time.

Struggling with our little manufacturing operation, we stayed on Abert Place for quite awhile, almost three years, mostly because we couldn't afford anything bigger or anything fancier. But there soon came a time when we were so crowded there was no way but to get a bigger place. A friend of mine, Tom Bentley, Sr., was in the commercial building business, and started to work on me to *build something*, but I would always say, "Tom, no damn way. We just don't have the money."

"Yeah, Mike, but you're totally out of room, and you're going to have to do something," Tom would say. "You're to the point where arms and legs are hanging out of the windows and doors there on Abert Place."

"Ah, maybe you're right, Tom, but I'll be damned if I can figure out how we're going to afford building something, even if we finance it up to the hilt," I said.

But somehow we did. We built a 4,800 square-foot brick building up on Elm Street, still in Milwaukee, but about two miles further north. Three times the space we had on Abert. We borrowed the down payment from the bank, and the mortgage payments were less than what we had been paying in rent.

The emotional impact was something else, however. It was a mixture of utter joy and pride coupled with a deep fear that kept me from sleeping at night. What if we didn't get any more business? What if our profits couldn't sustain the payments or we started to lose money? What if it turned out that we had built too big a building and my friends started kidding me about the white elephant I built? At times, I think I must have inherited my ability to worry from my dear mother, who was the *original* worrywart.

Once we sold the first central EKG system to Northwestern Medical School, we heard that a cardiologist by the name of Stuart Reid from Montreal General in Canada was very interested in a similar system, so I rushed up to see him.

A nicer man I have never met. He was indeed wanting to do something spectacular with his EKG department, and a system like Northwestern's was just the ticket, or so I told him. Dr. Reid finally ordered, but it took him what I thought was an infinity to make up his mind. I remember sweating bullets waiting for the order because we really needed more work back in Milwaukee. Since the original delivery at Northwestern, we had taken on all sorts of miscellaneous work, like some Automatic Anodizers for Delco in Indiana. But that's just what it was – miscellaneous. The T-BASS electronic bass fiddle was dead.

Our EKG system solved a number of problems with electro-cardiology in big hospitals, however, and a number of hospitals around the country were interested. First, it got rid of what was called cutting and mounting, a process where paper strips, recorded at the patient's bedside, were brought down to the heart station, cut and glued on a big piece of cardboard, and then presented to the doctor for reading.

The way our system worked, the tech would go to the bedside with our special machine on a trolley, hook up the patient and then plug into the phone jack. The signals would then be sent over the phone line to the heart station where they'd be automatically recorded and come out in a form ready for the doctor to read directly.

The systems were expensive for those days ($50,000) but saved a whole lot of labor and could easily be justified to the administration of a big institution. After Montreal General came orders from Johns Hopkins, Massachusetts General and then New York-Cornell. About thirty-five very large hospitals eventually bought our system. We made about one every month, and for awhile we got enough orders to keep us very busy.

Then, suddenly our luck ran out. The next possible order we counted on was being held up for some reason or another, and we couldn't seem to push any new orders through. I remember literally begging some of our potential buyers, but to no avail.

We had no choice but to close down or find something else to make. We scurried around both inside and outside the medical profession and finally found a project at Yerkes Observatory on Lake Geneva that we thought we might be able to build. It was a possible order for a one of a kind device called a spectrum analyzer to help astronomers determine what stars are made of. We had no experience in this field, but it was work and that was all that mattered.

After a long while, the central EKG business began to pick up again, not forever, mind you, but at least for awhile, and we were the only company making such a machine. We had very little money so we'd have to wait to get paid for one so we could afford parts to build the next system. Several times, we went into a total slump, but we always found some other system or device someone wanted and no one else was willing to build.

Northwestern's system in Chicago was working fairly well, but troubles with our equipment in other locations grew more serious by the day. Of the systems sold, one of the most important was at New York Hospital (Cornell University Medical School). There, problem after problem made my life miserable, not to mention how Cornell felt about it.

One spring day, as I was visiting New York-Cornell, I ran into my old friend, Lee Winston, a cardiologist who was more or less in charge of the heart station.

"Lunch?" he said.

"Sure, Lee, how the hell are you?"

"Great, Mike. Let's go to the Cornell Club. That way you won't have to pay 'cause it's a private club, and I know how cheap you are."

"Give me that, Lee. I always want to pay, but you never let me."

"Well," he said. "You always look like you're on your last dime, Mike. Besides, I owe you one. Come on, let's beat the crowd."

The Cornell Club was one of those wonderful old places for doctors only that you sometimes find in teaching hospitals, and this one was full of beautiful wood paneling and elegant china. The floors were random planked oak, and the place had a genuine Early American aura that made guests feel privileged to be there.

Lee said hello to a number of his fellow physicians in their white coats and made sure I met them as we moved through the crowded room.

Soon we were at our table, and Lee ordered ice tea for both of us. I would have rather had an Early Times, but no booze in this place. A pleasant ray of sunshine poured through the window where we were seated.

"How's the system working, Lee?" I asked, hoping for the best.

"Just fine, Mike. Things are going well. Really, no complaints. How's that chick you're chasing down here?"

"Oh, she's great, but it's no big deal, you know, Lee. I mean I'm not about to get divorced again. Does tend to get me to New York more often though."

"Ah, Mike, I thought the only reason you came to New York was to see me and make sure our system was working properly."

"It is, Lee, but this girl sort of makes it even more fun to come here. System really OK?"

"Absolutely, Mike. You know I'd be the first to let you know if we were having problems. By the way, there is one thing that's sort of, well, strange that shows up every once in awhile with the patients."

"What's that, Lee?" I said with some trepidation, hoping it was nothing serious.

"Yeah, well every now and then, some patients complain about a sort of tingling sensation when we connect them up for an EKG, and I never know if it's in their heads or just their imagination with all those wires hooked to them." He was rather serious for Lee.

"Impossible for it to be our equipment," I said confidently. "Why, we have our leakage current down to less than a milli-ampere now, and I'd safely say you couldn't feel anything less than ten times that amount even if you were totally neurotic."

"Probably the patients, then. More ice tea?" he said.

Some years later, the Peter Bent Brigham Hospital in Boston discovered that they could fibrillate a dog with 20 *microamperes* of leakage current, fifty times less than what our equipment was putting out. In other words, we could, possibly, have electro-cuted a number of patients with this equipment we were so proud of.

We modified our systems to meet the new standard as quickly as possible. Today, I think it's safe to say that the FDA would never let that happen, but you have to realize that this was all new technology, and we just didn't know the dangers. Eindhoven, the Dutch inventor of the test, wired a patient up directly to telephone wires back in 1906 so he could record the patient's EKG some six miles from his laboratory. Just imagine if even the slightest lightening storm had come near those wires.

Another incident at New York-Cornell Hospital bears mention. It was only a few months after my luncheon with the good Dr. Winston and the *slight tingle* he said patients were experiencing when hooked up to our equipment.

I arrived for one of my usual equipment checkup visits at the hospital around ten in the morning on a gray and dreary day in good old New York City. As I spun through the glass revolving door, just off East 68th Street, I noticed a big fire truck parked out in front and some firemen hurrying through the entrance.

"Excuse me," I said, as a very large guy with the usual red fireman's hat pushed me aside. Unruffled, I proceeded to the third floor via the stairwell since the elevators in that hospital took forever to produce any kind of vertical transportation.

As I opened the door onto the third floor, I found myself in what seemed to be some sort of war zone. People of every shape and size were pushing through the hall just ahead of a hospital guard who was barking emergency directives.

"Move straight down this hall and to the wing at the left, ladies and gentlemen. There you will see a sign for the emergency stairwell and proceed, as quickly as possible, to the next floor down. Those of you in wheelchairs, please wait at the exit for assistance. Do not panic."

"What is this?" I asked in a loud voice while almost being trampled by the passing crowd. To my horror, I realized they were coming from the direction where our equipment resided. Turning upstream against the crowd, I began to make my way toward the heart station and saw a nurse I knew (quite well, I might add) and hollered, "What the hell is going on, Gloria?"

"Your camera, Mike, it's belching black smoke, and they're evacuating the place!"

"Come on, doll, I don't believe it. Show me what's happening in there, quick!" I screamed.

Slowly, we made our way to the heart station and, sure enough, there was our microfilm camera sizzling away and making ominous sputtering sounds as a curious fireman approached with ax in hand.

"Stop!" I hollered. "It'll get worse if you do that. Let me kill the main power. That'll stop the smoking instantly."

I hurried to the main switch and made a sort of theatrical gesture in turning off the power so the fireman would be convinced the danger was over and would spare the ax.

And he did. The smoke subsided, and he and the other firemen were now preoccupied with a window to let some fresh air in the room.

But before I noticed, my friend Gloria had filled a paper cup with water and was emptying its contents on the machine.

Of course more smoke belched from the camera again and the fireman with the hatchet moved in for the kill. SMASH, and he split the smoking machine apart with a resounding blow.

"STOP!" I screamed, but it was too late. Our friendly fireman had now reverted back to his basic training days and would not stop chopping until the last puff of smoke was snuffed out and our machine totally destroyed.

Repair was painful, but explaining the incident to the administration was much worse. We swore to them that this had never happened before and never would happen again. We agreed to replace the entire camera unit with a brand new one. We agreed that if anything like that ever happened again with our equipment, we would remove the entire system and be ready to face a tremendous lawsuit that would, they assured us, destroy our company forever.

CHAPTER 7

OUT OF MONEY, AGAIN

Marquette was started, as I have said, with a mere $15,000. Of course when the business started to take off we ran out of money. Surprise, surprise! Confronted with our first do-or-die situation, I managed to wrestle up another $18,000 to put into the company, but that was not to last very long. Soon we needed *big money*, meaning something like two or three hundred thousand dollars. Our local banker was at our doorstep every other day to "get some more substance into the company."

We were on the brink, and I emphasize this because so many start up companies I have since known, reach this point in their history and capitulate way too soon by bringing in outside capital. It seems so logical when everyone around you turns up the heat. They talk of dangerous debt-to-equity ratios and "losing it all" if something isn't done. And the venture capitalists stand by like vultures, rubbing their hands with glee, reasoning that it's just a matter of time.

The early 1970s were a most critical time in our history. It was a time for our budding company when moving too quickly to outside capital of any kind could have been a disastrous mistake. It might not only have prompted our loss of control, but an erosive process that might have been very hard, perhaps impossible, to stop. It could have been the turning point as to whether

Fred Luber

or not we were masters of our destiny or forever working for someone else.

At that point, my friend, Fred Luber, convinced me to change banks and introduced me to the people from the Marshall and Ilsley bank in downtown Milwaukee. We established a much better rapport with this group. Their president, Jack Puelicher, stretched the banking rules as far as they could be stretched, I guess because he loved entrepreneurs and believed in us.

But the problem of money persisted, and everywhere the drums of outside investors and venture capitalists were beating with a deafening roar. I had this almost frantic feeling inside that we would be swallowed up by powers that I couldn't control. I had been warned and warned, but decided, almost blindly, that we would go it alone, somehow. We would *sell* the bank that prosperity was just around the corner.

By now, Mayo Clinic had become a good customer of ours. Two doctors from that institution, Ralph Zitnik (originally from Northwestern University) and Ralph Smith, along with myself, decided we would, somehow, scrape together $300,000. As it turned out, that was the last penny that ever went into the company (until we went public many years later). It was a total last resort, a go for broke effort. If this didn't make it, we'd throw in the towel.

And somehow, with the additional money, we turned the corner. In 1969-70, we logged in $1,842,942 in sales, up 111 percent from the previous year. Of course with these small numbers, it's easy to increase sales percentages, but it did represent real progress. We also logged a profit for the third straight year in spite of *recessionary forces* as I wrote in my annual report.

In 1970-71, we managed to grow 12 percent in spite of a continuing recession, and the following year we knocked the cover off the ball with sales of $4,360,869 and a profit (after taxes), of a spectacular $312,647.

All this growth caused a mixed bag of happiness and panic because it created a further shortage of cash. In my letter to the stockholders for the year 1972-73, I wrote, in part:

> We have just finished another fiscal year of outstanding performance at Marquette Electronics. Sales were up from $2,910,286 to $4,360,869, an increase of 50 percent over the previous year, and profits rose from $163,971 to $312,647, an increase of 91 percent. We expect continued growth this year with sales over $6,000,000.
>
> We have completed our new expanded facilities for the circuit board department at our manufacturing plant and have started plans to move engineering to 3642 West Elm, a completely separate building nearby. This will then bring about another reshuffle at headquarters to provide sales, service and accounting with the additional space they will require in the coming years. Also, negotiations are underway to acquire seven thousand more square feet behind the manufacturing plant at 3820 West Elm Street. No plans are

being considered at this time, however, for the construction of a new plant – the thought being that Elm Street can house us for many years, and at a very reasonable cost.

With little more than a year's exposure to the market, our new Magnetic Data-Card™ System for the storage and retrieval of electrocardiograms can be called an unquestionable success. Systems have been, or are being installed in the nation's best-known medical institutions, and performance reports thus far seem gratifying, indeed.

Last February, Marquette announced a new electrocardiograph, the Series 3000, with a Pressure-ink Paper Writing System to totally eliminate the necessity of mounting EKG records, and to produce tracings of far better quality than any competitive machine. Several other new products were introduced during this year, and many more are scheduled for this year. We also continue to pursue totally new areas of medical electronics with an aim toward diversification.

We have spent much time in the past year studying the various ramifications of a public offering of our stock. Our conclusion is the same as it always was – that a public offering would be appropriate at some time in the next year or so. With the market performance of recent months, however, it is essential that we maintain complete flexibility relative to timing of such an offer, and your management continues to arrange our working capital accordingly.

Here, one can sense a bit of an undertow for a public offering, yet we decided to hold our own, at least for a while. The truth was that we were nowhere near big enough to consider such a thought.

But evident in my writing, innovations were being introduced frequently. In fact, it seemed that every time things started
to get a little slow, we'd come up with another product, and just
in time to save us from wilting.

The annual reports also began to feature a graph of sales from
the beginning.

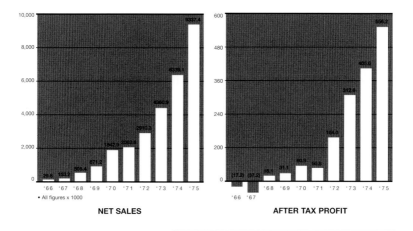

Marquette Electronics Sales and Profits 1966-1975

Meanwhile, Smith at Mayo Clinic, with the help of a group of
engineers from IBM headed by a PhD named Clyde Hyde, had
developed a computer analysis program for electrocardiograms. It
was the beginning of so-called *artificial intelligence*, and now
everyone getting an EKG at Mayo had his or her heart signals
processed by computer. It was a new and exciting way to analyze
the heart, and Smith's hope was to improve the accuracy and
speed of this old and valuable clinical procedure.

Smith was busy as a beaver with all these activities as well as
being a full-time cardiologist and running the EKG department
at Mayo Clinic. Zitnik, on the other hand, had gone on to greater

things. He had moved to Chicago in 1972 to head a cardiology service organization that would serve The Little Company of Mary Hospital and was to provide all of the hospital's needs in cardiology. His new organization was to provide EKG services for other hospitals in the Chicago area as well, and Zitnik could see big money in offering recording, storage, and interpretation of electrocardiograms throughout the area. He also could see the advantages of processing the tests by computer and so was very interested in getting a copy of the computer program that his old buddy, Ralph Smith, had developed at Mayo.

In fact, Zitnik lay claim to at least some of the program himself, saying that he was instrumental in its development while he was at Mayo. He asked Smith for a copy shortly after moving to Chicago, but Smith would have no part of it. Smith did not agree that Zitnik had any claim – rather that it was proprietary to Mayo. Besides that, it was still quite experimental, in Smith's conservative view.

The fur flew. Zitnik was livid while Smith dug in and refused to budge. I was right smack in the middle of the dispute although I had little to gain either way. Zitnik kept calling me to see if I could prevail on Smith. Finally, I talked Smith into giving Zitnik an older version of the program with a lot of restrictions regarding its use. Smith stalled and stalled but finally sent a reel of computer tape containing the older program.

Shortly after that traumatic experience, I was in Chicago and called Zitnik. "We'd better have lunch," he snorted. Zitnik reminded me a lot of a cartoon character popular in those days called "Senator Snort." He was a fairly big man in all dimensions and was an avid cigar smoker. We agreed to meet at the Mid-Day Club where all of the Chicago big shots have met for years. Zitnik was a member.

"How's she go, Zitch?" I opened with a friendly tone. "I'll bet you are one happy guy now that you've got Smith's program in your hot little hand."

"About time," he said, puffing his stogie. "And now I hear it's an older version the bastard sent me."

"Don't knock success, Zitch," I answered, sipping my big glass of Early Times. "I trust you have it stashed in a safe place, old boy."

"You bet I have," blowing a cloud of foul cigar smoke my way. "Got it in my brother's big, old steel safe 'til I get my programmer to load it on my CPU!"

Sometimes, I guess, I have a sort of twisted sense of humor because before I could stop myself, I said, "Old steel safe! God, Zitch, those old boxes are loaded with stray magnetic fields. I don't know, but I'll bet that environment could raise hell with your program on the tape."

Puffing frantically on his stogie now, Zitnik muttered several profanities, got up and shot out of the Mid-Day Club.

"Wait a minute, Zitch, I was only kidding. No way could that safe ruin the tape. Wait!" But it was too late. Zitnick was already out the door, scurrying down the hall as fast as his body would take him to his brother's office.

Magnetic tape is manufactured in a big *web* several feet wide and then slit by super-sharp knives to the desired width for various uses. Unfortunately, much as the manufacturer watches the process, these knives get dull occasionally, causing the tape to have a sort of fluted edge, making recording on it difficult.

How could fate deal Zitnik such a blow? How, after my complete buffoon about his brother's steel safe, could the very tape Smith sent him have had this fluting problem? Smith, of course, didn't know. The tape was undoubtedly recorded and never played back at Mayo. Only when Zitnik retrieved his precious taped program from his brother and tried to process it did he discover that it would not read.

Painfully, meticulously and agonizingly, Zitnik and his programmer worked their way through the defective tape, reconstructing the missing *parity check bits* inch by inch until the program was retrieved.

Zitnik forever blamed his brother's old steel safe.

CHAPTER 8

A LITTLE PIECE OF GE

Growth continued, again with the help of new innovations we introduced. By 1975 we had reached the dazzling sales level of $9.34 million, up from the previous year by 47 percent, and profits were over one-half million dollars. In my annual report to stockholders, I was happy, but showed concern about the world ahead of us, and rightly so, for the next two years were flat and agonizing.

Yet we had established a strong name in the field of medicine and continued to cut into the territory of big companies such as Hewlett-Packard. At first, HP scoffed at our equipment, saying, "Marquette makes nothing but fancy toys for rich doctors." But as time went on, all of our competitors began to take notice.

In the 1976-1977 report, we showed no growth for three consecutive years, but in my letter there was a little sign of optimism:

> Fiscal 1976-1977 again fell short of our ambitious goals, and for the third year in our history, we were without growth. Sales were $9,526,700 as compared with $9,734,500 last year, and profits were down sharply at $253,000 compared with $508,200 in fiscal 1975-1976.
>
> Our products are universally accepted by the medical field and probably the most advanced in this industry, yet

capital expenditures continue to stagnate as more and more pressures are brought to bear on the high cost of medical care. The future for us lies in absolute proof of cost reduction to the hospital or clinic using our equipment.

Profits, of course, suffered because of our reluctance to cut back on our overhead, always anticipating better sales tomorrow. Yet in the most recent months, particularly since year's end, incoming orders are up sharply, and look as if they may continue in this direction. Since we have no way of really knowing this, however, we have adopted a budget plan for the new year whereby we will make necessary corrections at the end of each quarter if orders and shipments do not produce the desired goals for profit."

And there was good reason for optimism, because in 1977-78 sales shot up by some 23 percent.

Fiscal 1977-1978 was a record year for Marquette, both in sales and profits. It was a good year, too, in the fact that for the first time in our history we carefully planned our course for the year, and met or exceeded goals in all respects.

Sales were $11,707,100, up 23 percent from the previous year, and profits were $855,000, representing 7.3 percent of sales. Performance records were mostly a result of careful planning and hard work, but some year's-end reassessment of our European accounting procedures also helped the bottom-line success.

During the latter part of the year we installed eleven of our new MUSE EKG Storage, Analysis and Management Systems in some of the most prestigious hospitals and clinics in the world. The future of this new product, indeed, looks promising since it is the most advanced of all systems competing for a market which could reach one to two hundred systems per year.

Our CASE Computerized Stress Test System continues to lead in technology and sales on an international basis, and the Series 2000, 3000 and 4000 electrocardiographs now produce almost one-quarter of all EKGs recorded in the United States.

In 1982, we built a Research Center near the main plant to see if we could stimulate more innovations, our life's blood. That same year I received a phone call from Walter Robb, then president of GE Medical.

Dr. Robb is a tall, slender guy of about my age and a brain chuck full of knowledge in chemistry, electronics, imaging, and particularly magnetic resonance imaging (MRI). If the truth be known, Walt was the driving force at GE that brought about the popular use of MRIs in most modern hospitals.

Having moved out from Schenectady, New York in the early eighties to head the Medical Division of GE at Waukesha, Wisconsin, Walt trans-formed the operation from ho-hum to an electrifying entity in which Jack Welch, the famous chairman, must have taken pride.

Out of the blue one day in 1982, Walt called me and asked if I'd like to have lunch at the University Club. I had never met Walt but was delighted to

Walter Robb, President of GE Medical

accept, although I didn't have a clue as to what might be on his mind. After lunch, he confessed that he wasn't a member of the club so I'd have to pay the bill.

One part of his operation that was somewhat of a problem for Walt was the monitoring product line. It was suffering under intense competitive pressure from Hewlett-Packard and another, smaller company called Spacelabs Medical. GE was really low man on the totem pole in monitoring (devices used to keep track of the vital signs of critically ill patients), and Walt had the idea that if GE could buy Marquette he'd reach a more *critical mass* and, perhaps, stop losing money on what was a bit of a thorn in the side of the Medical Division.

After a very dry martini, Walt began his sales pitch about how much better off we'd be if we were a part of powerful GE. I was very impressed with Walt, but unmoved by the idea of selling out, especially to a huge company like GE. I had recently written a little piece called the *Sellout Syndrome* in which I took issue with small companies that sell out to big business just for the sake of a buck (*see page 85*). I asked Walt if he had ever seen this epistle, but he said he had not.

A little hurt, I told him that it had been published in a number of trade journals and I was surprised that he hadn't come across it. I said there was no way I'd consider selling out at any price, and he'd understand if he read my *editorial*. Walt said "at least think about it, will you."

That night, I called my old pal and board member, Fred Luber, and told him about the discussion with Walt.

"What did you tell him?" asked Luber. "I suppose you gave him that 'sellout syndrome' stuff you've been preaching! We'd better get together and talk. As a board member, I really should know all about it."

Sellout Syndrome

'Never' is a long time and so that word probably shouldn't be used. But I'm tempted to say it in the face of so many other medical electronic firms selling out all around us. I cannot believe that money really means that much.

In 1965 I started Marquette Electronics in a funny old building that had 1600 square feet of space. I invested $7,500 and my partner another $7,500. Life was an unbelievable struggle. Designs didn't work, the customer didn't pay, the money ran short, and I worked fifteen hours a day. My partner continued with another business we had so we could both eat.

Today Marquette is $95 million in sales, 270,000 square feet of modern manufacturing facilities, some of the best researchers in the business, and world leadership in electrocardiography.

Every week someone contacts me to ask if we would sell out, be acquired, merge or otherwise mess up our beautiful world of success. People suggest absolutely wild prices, unbelievable benefits, and imply that if we won't budge, we'll be destroyed by the "giants."

Every month some company like ours is sold for those fancy prices to some overly profitable giant, and every day I hear horrors from the guys who stick around for the bureaucratic onslaught. Every day I hear how the good old Yankee entrepreneurial spirit is broken into millions of pieces by a conglomerate without spirit, heart or even an ability to get things done.

Of course there are special circumstances that might necessitate a sell out. The key guy dies. Some bad luck drains the financial resources dry. A competitor strikes a tough blow or the government legislates you to your knees. And, yes, I know I can't be the chief executive forever, but does that mean we must sell out and destroy what we have struggled to build over the most productive years of our lives?

These would be buyers tell me we need more money to keep up with research. Hog wash! I can't pump money into our Engineering Department of brilliant guys to make them invent or develop faster. We always seem to be ahead of our competition in innovation anyway, so how can more money make it happen better or faster?

They say we'll need money for continued growth. But our profits provide for our growth, and we grow 15-25 percent a year and make sure to stay profitable so our bankers loan us the needed operating capital. They said we'd need money for new facilities, yet our beautiful 270,000 square foot plant was entirely financed by an industrial bond issue.

So what, then, are the alternatives? Just keep growing and going forever? We all get old in the end and must make room for the young if we wish to perpetuate our company and what it stands for. If we have kids that want to step in, that's fine, but they have to want to and they have to be every bit as talented and ambitious as anyone else.

Our government has worked hard for many years to make the transfer of ownership in a company like ours to our young talent almost impossible. They have set capital gains taxes, gift taxes, income taxes and miscellaneous taxes of all kinds such that it's easy to go broke or go to jail if you ever get such a dumb idea, but we're going to try anyway. We're going to set our course for the next ten years so that we let the young, talented, capable guys in our company become major stockholders. And the stock will come from us... the old timers who must step back and have faith that someone else can carry on the work we're doing to advance the science of electrocardiography.

Someday I may regret writing these words. Someday I may wish we didn't turn our backs on those big juicy offers from conglomerates, but I don't think so. And in these words may be a lesson to others about what made this country and its industry great in the first place – and it surely wasn't going part way and then selling out to a devouring giant for the sake of money.

Sincerely,

Michael S. Cudahy

Michael J. Cudahy
President, Marquette Electronics, Inc.

The next day Fred and I got together and talked. I didn't really see what there was to discuss, but Fred had other ideas.

"Why don't you buy out GE's monitoring product line, Michael?"

"Are you nuts or what, Luber? Have you been smoking something? Why they'd never sell it to a little pip-squeak like us!"

"Don't be so sure, old friend. I hear they're losing a million dollars a month on the operation. Now if you were Walt, wouldn't you consider *any* solution to that?"

"Freddie, even if Walt would entertain such a mad scheme, he'd want umpteen gillion dollars, and we don't have a dime to spare."

"Not so fast, Michael. You could offer to give them convertible debentures. That way it wouldn't cost you any cash."

"Now, Fred," I said, "What the hell are convertible debentures?"

"Well, they are — well, sort of convertible preferred stock. At some point GE could convert them to ordinary stock."

"OK, Fred. You and your harebrained ideas. Why don't you check with Mel Newman (the company lawyer) or our accountants. Not that you're not a leading expert on such matters, but I'd love to hear what they have to say. Besides, do we really want to plunge into patient monitoring? I love adventure, old boy, but that's a hell of a leap."

"Michael, you are getting older, aren't you? Why, a year or two ago you would have leaped at such a chance."

"Later, Fred! I'll wait for Mel's comments."

But Mel thought it was an interesting idea. So did many others in the company. Old Warren Cozzens, our co-founder, was a bit skeptical, but soon came around as did our accountants. GE offered to let us have any and all of their employees that were within that group and willing to move. They sponsored a huge gathering of eligible people way over in the Hawaiian Islands so we could get to know them and recruit. I made an impassioned speech and promised we'd be the number one monitoring company within five years.

We did it, but it took until 1998.

About 250 people were engaged in GE's Monitoring Product Line, but only seventy-two decided to take the plunge, and that turned out to be a saving grace. Within six months we were showing a profit in what we called the Monitoring Division so it seemed that GE had way too many people for a gross volume of $25 million.

Unfortunately, we had too few with only seventy-two. In short order we found we needed a lot of redesigning of the products to make them capable of taking on the competition head-on. We tried to talk a few more GEers into moving, but our unstructured management style frightened a lot of them away, particularly engineers. One guy told me he just couldn't drive to work without a clear understanding of exactly what was expected of him before he started the day.

In 1984 I wrote to the stockholders:

During the year we moved vigorously into the patient monitoring business, determined to become a leader in this field by innovation and technological superiority. The medical field is beginning to recognize our contributions and seems to be rejoicing with our challenge to competition. Telemetry of a very revolutionary, advanced type has been added to our monitoring line.

A full line of newly developed operating room equipment has emerged from our research laboratories, as previously noted, offering features found on no other products. Cooperative efforts with the Medical College of Wisconsin have been fruitful in this regard, especially through the use of our jointly constructed operating room.

In 1985, under *Research and Engineering* I wrote:

Over 8 percent of our net volume is reinvested in research and engineering. I feel very strongly that this is the way to continued growth and success. As I said, we are already the leaders in many areas of medical electronics and determined to be in others. The only way to get there, in my book, is to be technologically superior.

In the past year we have devised a new fiber-optic blood pressure transducer, a precision, low cost, high speed modem for EKG telephone transmission, a remote access module for operating room monitoring, a capnograph for the operating room and intensive care, a new multilead arrhythmia software package for coronary care monitoring, a superminiature computerized electrocardiograph and new digital writer/printer to use with many of our products. Our new patient telemetry systems represent a substantial advancement over any previously available product from any other company.

The next year, we proudly announced the TRAM, a new monitoring innovation, which stole many a major account away from HP. By then it was getting more and more obvious to our competitors that Marquette was a force to be reckoned with.

Again from the annual report:

Fiscal 1985/1986 again set new records in sales for Marquette. Our total volume was $91.9 million, up 18 percent from the previous year. Before-tax profits were also up from $10.9 million to $13.2 million, an increase of about 21 percent.

The following year, I wrote:

> Marquette has just completed a year of growth that will be hard to match. Sales were up 35 percent at $144.8 million. Profits, too, were up considerably at $16 million after taxes.
>
> As predicted, Monitoring Division sales surged ahead almost 100 percent, resulting in good profits, as they became one of the world's leading manufacturers of such equipment. The Diagnostics, Service and Supplies Divisions also did extremely well.
>
> International sales jumped some 46 percent over last year, partially because, I'm sure, of a more realistic value of the dollar, but also because of our vigorous marketing efforts and the fact that we now are fully recognized in so many countries as a leader with special technologies.

ACQUISITION

As we finished our fiscal year, we also completed the acquisition of Temtech, a defibrillator firm located in Northern Ireland. Temtech has been renamed Marquette N. Ireland, Ltd. and will be operated as a separate division. With advanced technologies in this important segment of medical electronics, many of our products will be considerably enhanced by defibrillators. U.S., as well as world markets, will give Marquette N. Ireland a unique opportunity for growth with the addition of Marquette's national and international distribution network.

REAL ESTATE

We have completed an agreement to purchase sixteen acres of land in Jupiter, Florida (about twenty miles north of Palm Beach), and the Board will soon consider the possibility of building a plant there for the Supplies and Service Divisions. When, and if, this move occurs, we will probably expand Plant II on Tower Avenue in order to accommodate the entire Diagnostics Division, giving the Monitoring Division more room in Plant I. All of this could possibly take place in the time frame of our next fiscal year.

RESEARCH AND ENGINEERING

More than $11 million was spent in the past year to assure our company of continued technological excellence. The TRAM™ (Transport Remote Acquisition Monitor) systems are being refined in clinical use and have been accepted with great enthusiasm by many prominent medical institutions. Electrocardiographs are being introduced with precision pacemaker analysis capabilities and our newest versions will record and analyze minute *after potentials* of the post-heart attack patient, hopefully giving clues as to the possibility of further complications before they occur.

Marquette was on a roll with even more rapid growth than in the previous years and innovations were being announced every few months. Competition was going nuts!

CHAPTER 9

THE MARQUETTE CULTURE

Back in the eighties I wrote a one-pager called *Cudahy's Ten Golden Rules on How to Run an Organization (see page 92).* It was a bit of a buffoon, but with everyone else in Corporate America ensconced in Company Rules, I thought I'd make a bit of a joke out of the whole subject.

To my surprise, however, it caught people's eyes when it showed up in various newspapers. I must have touched a nerve end or two with things such as, "Don't have meetings – they're the curse of American business" or "Ignore the competition – just do your job" or "Never make an organizational chart – it puts people in boxes." I got letters from all over the place, believe it or not, and I'm forever running into someone who has it posted on their office wall. I'm flattered, to say the least.

Sometime after I wrote *The Ten Rules*, I read a book called *Further Up The Organization* by Bob Townsend. He gives a quiz on management philosophies and

Bob Townsend, author of
Further Up The Organization

CUDAHY'S TEN GOLDEN RULES
ON HOW TO RUN AN ORGANIZATION

1. **Stay financially independent.** Don't get the venture guys in too soon no matter how difficult it is.

2. **Don't have meetings.** Meetings are the curse of modern business. Most break up without a decision. If you must get together, do it in the parking lot or have no chairs in the meeting room.

3. **Never make an organizational chart.** Putting people in boxes always makes some smug and others mad. A true representation of how people work together would look like a Picasso painting and then, of course, no one would understand it.

4. **Promote from within.** This one Laurence Peter said twenty years ago, but it deserves repeating. Obviously, if your company is growing fast, it's hard to do, but if you're continuously going to the outside for talent, you're probably doing something wrong.

5. **Don't hire consultants** every time you have a problem. Consultants are OK if you don't know how to perform some complex process, but most problems, like quality or efficiency or inventory, are things you ought to solve from within.

6. **Have a love affair with your employees.** I mean all of them. Trust them, embrace them, tell them your secrets and treat them like part of your family. This is a good way to treat your customers too.

7. **Give away the store to your employees.** Sure, they have to earn it some way, but most bosses lean way too far the other way. I have optioned, granted, sold and otherwise given a gigantic part of my company to employees from top to bottom, and it's come back to me a hundred times over.

8. **Don't play big shot.** No reserved parking or private entrances. No fancy offices and thicker carpets for a chosen few. And let the people do what they want in their own area. Crazy posters, radios, plants and decorations of any kind. It'll make the place home, and people show up for work a lot more often. Let peer pressure prevent anything outrageous.

9. **Ignore the competition.** More people waste more time worrying about what the competition is doing than they do tending to their own knitting. Just keep a steady course in the direction you feel is best for your customers' needs. Elbow grease is a good substitute for paranoia.

10. **Finally, preserve your sense of humor** . . . at all costs. Sure, times get tough, but if business isn't fun anymore, give it up and become a cab driver or bartender. You only live once.

Michael J. Cudahy

then says, "If you get a score of nine out of ten, I'd like to come visit you."

I took the quiz and scored nine, so I Xeroxed the thing and sent it off to him along with my ten rules. I got a quick response saying, "When can I come out to see you?"

When Townsend arrived, the *Milwaukee Journal* asked if they could record our comments for a feature article. Here's the newspaper transcript, in part, which took place on May 23, 1985. The reporter started by asking me if it was difficult for people to adjust to the informal structure of a firm like Marquette Electronics.

Cudahy: You can see my office. My office is the same as anybody else's.

I will never forget going to see [Thomas] Watson when he was chairman of the board of IBM. You started on the lowest level, then you got to the carpet level, and then came the bigger desks and the fancier offices and finally you got to Tom's office – and now your feet are disappearing in the nap of the carpet.

This kind of thing has just got to make people madder than hell that work in the back [at the manufacturing level]. As I have said many a time, if you want a fancy life and you've done well, you can have it at home."

Townsend: The guy that took over for me – I always used to kid him. I said, "You're not comfortable unless there's a crystal chandelier in your office." He was a guy that needed a lot of trappings. He needed lots of secretaries and assistants and things like that.

And he was a lawyer. I don't believe lawyers make good chief executive officers. It's a risk-averse profession; they tend to try to gather too much information. They also do not have faith in their fellow human beings, which makes them susceptible to headhunters [professional recruiters] when they go to fill a key position, and that's a terrible mistake to make.

If you fill a high-level position from the outside, think of it this way: You have three people in the company who thought they ought to have a shot at that job. They all have followers. So when this chief executive goes to a search firm and brings in an Electric Blue Suit at a high price and puts him in the job, he's offended all three of these important constituencies. He's turned them right off.

Cudahy: Absolutely. I agree with you 100 percent.

Reporter: Do you promote from within at Marquette Electronics?

Cudahy: It once in awhile becomes necessary to go outside, but only after you search and search and search and exhaust every possibility.

To be truthful with you, I made a prize mistake one time of hiring a top-notch guy from the competition. The only trouble is he was top-notch because about 96,000 people held him up there. He couldn't do anything.

The day I decided he had to go was the day that he and I walked through the plant, from one end to the other. I must have said hello to a hundred people, and he didn't say hello to anybody. I said, "Gee, do you realize you don't know any of those shop people?"

He said, "I don't need to know them."

And I said to myself, "And I don't need you."

In discussing my plans for this book with friends, the number one request was to make sure that I explain the culture of Marquette. Not easy, since it was never a planned culture. It just happened.

One day, I was having lunch with a group of guys who came from a large and structured company that will remain nameless. They were trying to get me to tell them how we operated, and I said things like the reason we never had time clocks was because we always trusted people. We assumed that almost everyone would put in a full day if they loved what they were doing and wanted to be a part of our success story. That simple. Besides, our employees all had a piece of the action in our ESOT plan (employee stock ownership trust) so success was theirs as well as mine.

I'm afraid the group was not quite buying all this. "What about those who *do* goof off?" they asked. "There always are a few who would take advantage."

"True," I said, "but that's probably less than 2 percent of all employees. Why have time clocks for the 2 percent and make the other 98 percent suffer?"

"You've got a point, Mike" they said, "but is that suffering? I mean, come on, what's the big deal about punching in?"

"No big deal, true," I said, "but when you have to punch in, it says that the management doesn't trust you, and that's not what I was trying to tell them – in fact, quite the opposite."

This whole thing about trust *was* a big deal, I told them. It said that if my employees were convinced I trusted them, they would perform in an entirely different fashion than most people normally do in a company that has time clocks.

These were, of course, just a few dimensions of Marquette we discussed that day, which, by the way, seemed to go in one ear and out the other. But in order to better understand the mysterious qualities related to the Marquette culture, let me offer six short stories.

Employees sign the slab for the expansion of Marquette's Tower Avenue facility

LE BISTRO

In 1983, we again found the need to expand our facilities at 8200 West Tower Avenue in Milwaukee, by at least 50,000 square feet. I was determined to make sure we would end up, this time, with the right amenities for our employees, such as a nice place to eat.

In this vein, we set out to design a restaurant that would seat two hundred seventy-five people and would have no resemblance to the typical company cafeteria. I had seen a thousand *lunch rooms* around the world and always felt they were depressing, demeaning, and demoralizing.

Why, I wondered, should this have to be? Typical features of the so-called *company cafeteria* are tan linoleum floors, cold metal chairs with plastic cushions and long, Formica tables that seat ten or twelve people. This provides little in the way of camaraderie, and absolutely no privacy. PA systems blare a page for someone wanted on the phone, and the place smells of stale French fries. Low-cost fluorescent lighting flickers with tired bulbs, casting a brownish green hue on the people below.

I swore I'd do something better for the troops. Much better.

So we built a really nice place and called it *Le Bistro*. We hired two good, young chef/manager guys, Aaron White and Bill Kuptz, and told them to make it work. Serve high-quality food

Cudahy drove the earthmover in 1983 for the ground breaking ceremony of a new plant addition.

at prices comparable to that found on the outside and make it reasonably profitable. We'd use the profits for a company picnic.

We constructed a full, elevated stage at one end of Le Bistro and equipped it with most of the lighting and other gear used for modest theater productions. We carpeted the floor, and I commissioned an artist friend to create ten big, bright watercolor paintings for the inside walls. Wood furniture, brass trim, and soft, incandescent lighting – it was unlike any other company cafeteria the world had ever seen except for the one at the Allen-Bradley Company. I stole the idea from there.

When the plant manager came around to ask me if we could connect the sound system with the company paging system, I said, "Hell, no! If you want to find someone in Le Bistro, you'll have to get off your duff and look for yourself. People shouldn't have to put up with that yacking while they're trying to relax and eat."

I'll admit we didn't really have anything specific in mind when we built the stage at one end. It was just that Allen-Bradley had a stage for their company jazz band, which sometimes played at lunch time, and I thought that idea was pretty cool.

Employees relaxing at Le Bistro

We also decided to serve beer and wine. I had been in many European factories that had served beer for well over a hundred years so I thought we couldn't be too far off base with the idea. It was another way to say to our employees that this was no ordinary factory. Besides, we reasoned that if a guy was going to get buzzed up over the lunch hour there were no less than a dozen saloons in the neighborhood.

It turned out we had absolutely no problem with the idea. In fact, the alcohol consumption per capita was surprisingly low. On Fridays, after a long, hard week, some groups of employees would gather in Le Bistro for a little partying, and that turned out to be a plus, having everyone go over things in a relaxed atmosphere before taking off for the weekend.

Of course the local newspaper had to do a story about this "booze in the workplace" thing but they did a pretty good job treating the subject fairly.

Business

MILWAUKEE SENTINEL

LYNN HOWELL
Sentinel photographer

Cudahy uses trust, perks to motivate Marquette workers

By AVRUM D. LANK
Business Editor

It's 9:30 on a Tuesday morning at the Marquette Electronics, Inc. headquarters and main plant on Milwaukee's Northwest Side.

Throughout the building there is an air of focused, managed intensity as approximately 750 people who work there begin their day.

In Le Bistro, the company's cafeteria just off the factory floor, the atmosphere is more relaxed.

As workers clean grills and start to lay out sandwiches for lunch, the smell of popcorn fills the air. Coffee bubbles in large urns on one wall.

A steady stream of employees – assemblers, research scientists, accountants, secretaries – saunters by, filling cups with coffee, chatting briefly and leaving.

Among those in the lunch room is a smiling, slightly built man with white hair and a bright smile.

He is completely relaxed, and his purple sweater and chino pants blend well with the casual attire of those around him.

His shirt is open at the throat. A neck tie is as rare as a faulty circuit board at Marquette, a maker of high-tech medical equipment.

The man is Michael J. Cudahy, president and co-founder of the private company, and the greetings and good-natured insults he exchanges with those around him belies his authority.

"Is this a typical pose?" asks one employee walking by as Cudahy, at the request of a newspaper photographer, is drawing some beer from a keg to one side of the room.

"Shut up, Paul," Cudahy answers. And both men chuckle.

Paul goes to the coffee urn and fills his cup. He leaves without paying, because, as Cudahy points out, the coffee is free.

The beer is not free, but it is there for anyone who wants some with lunch or dinner, as is wine, also sold at Le Bistro.

Marquette is probably the only company in the city to offer its employees alcoholic beverages daily.

Cudahy finds nothing unusual in the offer, however.

"I have been to a number of factories in Europe," Cudahy explained, "and they have been doing it for 50 years over there. It's just another example of how we trust our workers."

Milwaukee Sentinel *feature article on Marquette culture*

99

Not all readers of the article thought the idea of beer and wine at the plant was so good, however. One lady wrote me shortly after the article appeared and said:

I read this article in the *Milwaukee Sentinel* (12-13-88) about Marquette Electronics. I'm referring to the fact you not only allow your employees to consume beer and wine during lunch and dinner, you sell it to them. You were quoted in the article, "It is just another example of how we trust our workers."

I am both shocked and amazed that you, a businessman holding a high level position of responsibility, would exhibit an apparent total lack of knowledge and understanding of the alcohol problem facing this nation, a problem that is more predominant in Wisconsin than in most other states.

I like to be well informed when receiving medical treatment of this kind. For this reason, upon returning to Mayo Clinic, or going to any doctors or hospitals in Milwaukee, I will inquire if your equipment will be used. If the answer is yes I will refuse treatment and show them the *Sentinel* article. I will then tell them I would not knowingly let a doctor do a bypass surgery on me if he had a few beers. Why, then, would I be so foolish as to consider any high-tech medical equipment manufactured by Marquette Electronics.

Be assured, I will never forget or trust the name Cudahy or Marquette Electronics because of their policy on alcohol. I have obtained extra copies of the article and will be passing them on to hospitals and doctors.

With a full stage at one end of Le Bistro and a seating capacity of almost three hundred, my partner Warren Cozzens started to get ideas. He loved and often attended the legitimate theater in Chicago and New York. He was sure he was qualified as a playwright to create a show for Marquetters, and some of us who were really hams at heart were convinced we could all become actors and actresses if given half a chance.

He was right. In 1984, old Warren decided to write a sort of parody, *The First One Hundred Years*, about the company, even though we'd only been in existence for about twenty.

Everyone thought this was a fabulous idea and encouraged Warren to get on with it. They were sure it would be a smash hit if old WBC wrote it because everyone loved Warren. He had no trouble finding volunteers or would-be actors and actresses out of every corner of the company – everyone from VPs to maintenance people.

Of course there was some griping about lost production time with all the rehearsals that went on and also some jealousy since all applicants weren't accepted by our newly found *Cecil B. DeMille*. But, for the most part, they took the whole thing as one of Cozzens' and Cudahy's crazy ideas. Besides, it was near Christmas time and things were a little slower, factory production-wise.

The result – we had a packed house for four great performances of *The First One Hundred Years* over the holidays. The audience, consisting mostly of our employees and their families, cried and laughed and cheered through every show. When it was all over, we found a new and even closer bond with every person in the outfit – in the show or not. We realized we probably couldn't pull off such a thing every year, but for the time being, it created an electric moment for the company. And Warren Cozzens was our hero.

We had two more shows in the years to come. Each one got progressively better, or so we thought. The best was *The Fountain of Youth* and consisted of a story about how one of our computer whiz kids had invented a time machine that could alter the course of history.

As the play opens, our engineering genius is seen on stage tinkering with his time machine. As a result, he succeeds in changing the history of Marquette Electronics so it gets its start in Florida rather than in Milwaukee and becomes *Ponce de Leon Electronics*.

There were wonderful scenes of dancing and merriment. The acting and choreography was amazingly professional considering that the people were right out of the rank and file of the company with little or no acting or dancing experience.

One of the scenes, after the time machine had changed the history of the company, showed how poorly we were doing in the deep South with *Ponce de Leon Electronics*. As the curtain opened, only a few workers were left and they had taken to excessive drinking of a home-brewed elixir. I was one of the *stars* and sang a song I wrote for the play called the *Alligator Blues*.

Well, now down here in the Southland, I got th'
    alligator blues
Yea' down here in th' swampland, I got alligators snap-
    pin' at my shoes
There's nothin' down here for me, baby, 'cept my
    old Early Times booze

Now, I shoulda stayed in Milwaukee, I coulda' made my
    fortune there
Yea, if I'd stayed in Milwaukee, I wouldn't had all this
    disappear
I coulda' had three wives, eleven children, and even had a
    little money left to spare

Cudahy sings the blues in The Fountain of Youth *at Le Bistro.*

The reference to three wives and eleven children, of course, was a standing joke about all my wives and children. And having the CEO up on stage in ragged old clothes, singing his heart out with this blues song sent the crowd into screams and roars of laughter and applause. At the end of the last performance, the stagehands obtained a live alligator from the local zoo and presented it to me at the curtain call. I tried to keep my sense of humor, but I'll admit I was terrified even though the poor critter had duct tape over his mouth.

Why did we ever give up this idea of having a play every Christmas? Unfortunately, partner Warren grew older, and the strain was just too much for him. Then, too, there were some managers bitching at all the time the production of plays took away from the production of medical products.

Marquette style show

Cudahy as a Halloween hooker

But these plays were not the only extracurricular activity at Marquette. We had fashion shows and going away parties and coming home parties (some people quit and then came back when they found out how the *real world* was). We had *Success of a New Product* parties and even *Failure* parties. Any excuse was good enough. Over the years, various groups would cook up a notion like a Halloween costume party and then just do it. One year at that party, someone got me to dress up as the town hooker. I won second prize against 30 other contestants, and no one knew who I was until it was all over.

Fortunately, I've saved a vault full of pictures, posters and videos, including all the Cozzens productions.

THE KIDS' DEPARTMENT

In 1982, my daughter Julie, pregnant with her first child, had a nice job at Marquette. One day she said "Pop, what am I going to do about my job when the baby comes?"

Being the kindly father that I was, I said, "That's your problem, kid."

"Come on, Pop, I'm serious! Why don't you start a day-care center right here in the plant? There's a ton of parents in this outfit that could really use a place like that. It's not just me."

"I'll think about it, Jules, but don't nag me. I've got enough on my mind."

But she kept nagging anyway until I finally said, "Listen, Jules, tell you what I'm willing to do. I'll build a *pilot* day-care for a few kids, and we'll see how it works, OK? If it does, I mean economically and all, maybe then we can consider a permanent place."

We found a corner of the plant where we could move things around a bit and staked out about 1,100 square feet for the *pilot* and calculated that we could accommodate a maximum of twelve kids. Then the fun started. We soon found out there were state regulations on such matters, and we'd have to comply. The walls had to be cement block for fire resistance, not just dry wall. Toilets had to be added and quite a few other expensive amenities were required.

OK, we'll do it. I had promised, so there was no turning back. Hang the cost, hire the teachers and let's go for it! Make the applications. Figure out the insurance. What a nightmare! But we stuck to it, and in no time, we had the first on-site company day-care center in all Wisconsin.

The Lieutenant Governor, Scott McCallum, and the TV stations came for the dedication, and we were off and running with eleven kids and one more expected. In a matter of months we had another ten on the waiting list, and from there on it was expand, rebuild, expand, rebuild.

Our financial guys were having a fit. What to charge? What do commercial day-care centers charge? Should the parents with younger tots pay more? It was more work to care for them. Should we build an outdoor play area, and then what regulations would we have to meet out there? If we had appointed a committee – one from human resources, one from finance, one from facilities, one to study insurance – I can promise you we never would have built this crazy thing. Fortunately, we *didn't* have a committee. We just did it.

Somehow, we managed to answer all those questions and many more. With the iron will of the staff and parents, we succeeded, and then some. Everyone chipped in to build all sorts of things for the play area. Slides, swings, tunnels, climbing bars, all made with loving hands and all decorated with the skills of our own artistic in-house talent. It was a delight to everyone,

Mike Cudahy and some friends at The Kids' Department, Marquette's day-care center

parents and all the other employees. Everyone was proud of it, especially me.

The teachers taught, the kids learned, and the parents could go in any time to see if their kids were OK. We eventually had infants in cribs, pre-kindergarten, kindergarten and what we called the "graduating class." Every year we dressed this class up in caps and gowns and had a big, official graduation with songs, music, cake, and cookies. The whole plant turned out, whether they had a kid in the school or not. The TV stations always covered it because it was so much fun to watch.

One day we decided the day-care operation, physically, was getting to be a mess. We had expanded one direction and then another, and it was not only interfering with the production department, but was very crowded and difficult for the teachers. At that point we had over 100 kids attending.

We got together with the facilities guys and said, "We need 10,000 square feet of clean space to start all over again with the

Cudahy and day-care children on Marquette grounds

day-care," and they went nuts. "Where, may I ask, do you propose to find that kind of space in this overcrowded plant?" wondered Frank Schmidt, the guy who would have to sacrifice space from the production floor.

"Hell, I don't know, Frank," I said, "but you always find a way around problems like that, so I'm sure you'll come through, old boy."

He did. We soon began clearing a large area in one corner of the plant and made room on the outside for a substantial playground.

As the construction work neared completion, I took a tour one day and asked the painters if the gray wall paint was a primer, and the guy said, "No, mister, this is it. Gray and beige throughout. That's what the specs say, and that's what we're doin'."

What a bummer! The place is going to look like a prison, I thought. Something has got to be done *immediately,* but to change now will delay the opening and run up the bill for sure. I had to think *fast.*

Suddenly, I had a totally wild idea. We had just hired a brilliant young industrial design engineer, Scott Robinson, to make our products look less like engineering prototypes. Why not talk him into respecifying the whole day-care? Let him use crazy, modern colors throughout. Doors to doorknobs. Walls to ceilings. And let anyone and everyone who thought they knew anything about art do murals for the hallways.

"Let me see if I've got this straight. You want me to do what?" asked Scott. "I already have a pretty full plate designing equipment around here, so what am I supposed to do, drop everything and do this day-care job?"

"Scott, think how much fun you'll have," I said. "And we'll put a big plaque at the entrance saying 'decor by Scott Robinson.' Your name in lights, Scott. Everyone in the plant and all the customers will know you did it. The TV stations, when they cover the opening of the new facility, will make a big thing about your work! You can do the whole thing on weekends so you can get your regular work out as well."

"Boy, Mr. Cudahy, I mean Mike, you must have been in sales before becoming president of this outfit. OK, I'll do it!"

The result of all this was that we not only had one of the largest and best day-care centers in all of Wisconsin, but one of the wildest, most colorful places in the world. The kids loved it and loved our manager, Sandy Matter and her devoted staff. It was the state's model for in-house industrial day-care.

arRHYTHMia

Somewhere in the late eighties or early nineties I was listening to a jazz record featuring six French singers – no instruments. Among other selections, they sang a Woody Herman tune called *Early Autumn* exactly like the sax section of his band used to play it. The group was called the Double Six of Paris. The words were in French.

I was electrified by the sound of six voices perfectly reproducing the tight harmony of that great band of the sixties, and it gave me an idea. Why not bring together some of our employees who thought they could sing and form a company singing sextet.

I had no problem finding volunteers. Everyone thought it was a really fun idea. A lot of people thought they could sing, but the problem then was how to pick the ones who really could – I mean read music and sing really complex harmony.

But some of the volunteers were musicians and many had played and sung in bands around town so we were not starting from scratch by a long shot. We had no rules about when to rehearse, when to sing and when to go back to work, but everyone

arRHYTHMia *(left to right) Joyce Lewandowski, Gary Christensen, Mike Tacke, Michelle Hynson, Julie Podjaski, David Nicolai*

who joined the group recognized they could easily upset the other employees if they spent too much time singing and didn't take care of their regular jobs.

They named the group *arRHYTHMia*, a rather corny takeoff on our serious business of dealing with the heart, but the name stuck and got to be well-known around Milwaukee, especially at Christmas time. They rehearsed and rehearsed and finally sounded very professional, making appearances at shopping malls, holiday parties and picnics around the area. We used to get wonderful fan mail from people praising the group and praising Marquette for adding an extra touch to the city.

Visiting customers and the like were often startled to hear songs bursting from the rehearsal studio when taking a tour of the plant.

"What is that all about?" they'd ask.

"Well, we have many extracurricular activities for our employees here at Marquette," I'd say. "We work hard, but we play hard as well."

Most customers, after recovering from the initial shock, were impressed and reasoned that happy employees probably meant better equipment. A few would gasp in disbelief but probably couldn't wait to get home and tell their fellow employees about what they saw at Marquette.

At this point, it is easy to ask what all this nonsense has to do with business. Were we running a company or a loony bin? A circus perhaps? But it all goes back to what I keep saying. The more employees love their job – and the place where they work – the more productive they'll be.

PREJUDICE

Prejudice was never taught in our house. It was sort of an unspoken creed that everyone was to be treated as an equal. At school, however, things were different. I went to a private school in Milwaukee where snobby kids would say nasty things about Jews and blacks. The teachers weren't quite as bad as the kids, however.

This, of course, was in the 1940s.

I sometimes wonder if I am somewhat prejudiced. I have a black friend who says everyone is a little that way – it's just a question of degree, and maybe he's right. Surely some of those snotty things that were said in school could have rubbed off on me to some extent. It's important, I think, to examine your soul about matters like that, especially if you're going to hire people.

I had a plant manager by the name of Bob Klunk. It seems like a hundred years ago, but it was 1967. Marquette employed all of about twelve people, and this guy was out looking for another assembler. We had a pretty good backlog for once.

Manager Klunk came to me one day and said he had found a really good guy for the job and did I mind if he was black? I said I didn't care if he was green as long as he could do the job and that kind of set the tone for our antidiscrimination policy from then on. No big deal about it, just pay no attention to race, creed, color or sex, but make sure people were qualified for the job.

Some years later, the Office of Equal Opportunity came in and told us we'd have to post a nondiscrimination policy with a letter in the front lobby, signed by me.

Of course I said "fine," but then this lady from OEO proceeded to tell me I had to pick a letter from her sample book of *standard letters* and then I could sign it.

I blew my top.

"Come on," I said, "You want me to sign a letter that I had nothing to do with writing?"

"Well, yes," she said. "We have to be sure that you state your policy clearly and concisely, Mr. Cudahy."

"How clear do you want it? It's my policy. If I can't write it so it's clear I shouldn't have this job, don't you think?"

"Well, this is what I've been instructed to do, Mr. Cudahy. We have a number of letters to choose from, and I'm sure you'll like at least one of them."

"Ridiculous!" I bellowed, "I'll tell you what. I'll write my own bloody letter and then if you don't like it we'll talk some more, OK?"

"Well, I'll have to ask my boss what to do if you refuse. I'll come back next Tuesday and we'll see about this thing," she said, with a bit of anger in her voice.

So I wrote my *position* letter right from the heart, beginning with, "When we started this company, long before the Office of Equal Opportunity, we took the logical position that we would hire on the basis of skills and qualification, and in no way would race, creed, color or sex ever play a part in our judgment."

The lady came back the next week, and I showed her the letter. She didn't like it one bit. She said it didn't have the right tone. It was *almost prejudice* in her opinion. I couldn't, for the life of me, figure out how she came to that opinion, but that's what she said, so I gave her a challenge.

"Tell you what I'll do. I'll take this letter to five of my employees and ask them if they see anything prejudiced about it. Then, if even *one* says it *is,* in *any way,* I'll use one of your stupid form letters, OK? And I'll pick five employees who could easily lean towards your opinion like two blacks, one white female, one Mexican and one Caucasian male."

"Mr. Cudahy, this is highly irregular, and I'll, again, have to go back to my boss, but I doubt that we want to set such a precedent."

Surprisingly, this lady's boss said, "Let Cudahy try it. He'll fail." But I didn't, and *my* letter went up in the lobby and has been there ever since.

MAPLE SYRUP

Somewhere in the 1980s, the fourteen acre lot behind our main plant in Milwaukee was for sale, and someone was getting very interested in buying it and building a foundry. We loved the woods and open spaces and sometimes had parades back there. So, none of us liked *that* idea very much. The thought of a smelly, gray old foundry with a big smokestack belching pollution on us made me sick. And our day-care center was on that side of the building. It would have looked right out on the foundry in place of a beautiful woods which contained some of the most magnificent trees you ever could imagine. One old oak was some three hundred years old, the experts said.

Well, I bought the land and that was the end of the foundry. We turned it into a fantastic park for everyone. Jogging trails, picnic tables and all sorts of things that didn't cost much, but made it a great place for anyone at Marquette to go and relax. Most of the work to clear the trails and build a gazebo was done by employees without a *committee* being formed or any real organization. It just happened, and the gang loved it. So did people from nearby companies, but that was OK with us. It allowed us to get to know our neighbors.

Then a guy by the name of Ed Boldt, who worked in our machine shop, came to me one day and told me all about the beautiful maple trees that were back there in what we eventually named Marquette Park and said we should tap these trees for maple syrup.

"OK, Ed, I don't care, but so what? You want us to go into the maple syrup business?"

"No, Mike, not that, but it could be a kind of a special thing each year. We could get some of the gang and maybe some kids from the day-care center and have a sort of maple tree tapping party. Then we could *cure* the stuff and have a special breakfast with pancakes in Le Bistro. What do you think?"

"Go for it, Ed. Sounds like a lot of fun, but it's going to be cold out there. You did say the *tapping* has to be done in March, didn't you? I'm not so sure you're going to get a lot of people to do that."

"My problem, Mike. I'll get the people out there. Don't you worry." And he did. Big crowds. Employees, their families, kids of all ages, every year for many years.

We'd all gather together and first prepare plastic bags with a metal frame as a sort of *catcher*. Then Ed would pick out a tree, and one of us would drill a hole in its side and pound in a *spout* with a hook on it and then attach the plastic bag.

We'd sing songs and have sodas and make sure the kids did a few of the trees on their own. A few days later, we'd all go out there again and collect the raw syrup and put it in this kind of *cooker* Ed built and watch the stuff thicken into a delicious gooey syrup.

The third part of the ritual was to have a breakfast in Le Bistro, where Bill Kuptz, our chief chef, would cook up a huge batch of pancakes for Ed's syrup. Our choral group, arRHYTHMia, would sing songs like *Oh What a Beautiful Morning* to complete the corny scene, and everyone would eat too much and almost fall asleep.

In 1995, when Barry Allen joined the company as president, fresh out of his presidency of Illinois Bell, he had a chance to see this whole procedure, and I thought he was going to blow a gasket. He tried to be a good sport about it, but "waste of time and money" was written all over his face.

THE ITALIAN DISASTER

Our free and easy ways, although a part of our culture, got us into a fair amount of trouble from time to time.

Take the orders for a computerized EKG system we got in 1981. The University of Napoli Hospital had a rather rigid chief of cardiology, who insisted that all the computer statements from the system be output in Italian.

Big job, but we were determined to get the order so we really had no choice but to put a software engineer on the task of entering thousands of statements in Italian. The university furnished the correct translations from the English so it wasn't a problem of getting the wording right. It was merely a job of putting it all into the program.

Don Brodnick, a new and very bright engineer we had just hired, got the job. Day and night, weekends and holidays, poor old Don labored away at his keyboard translating things like *left ventricular hypertrophy* into the Italian version until he was getting a little wacky, or so some of his fellow engineers reported.

One of the most annoying things Don ran into in this process was when the readout said "no statement found" – a sort of cop-out in the program. After typing this statement in Italian several hundred times (recognizing his work would have to go through the scrutiny of proofing by someone else in the outfit, later), Don decided to put a dash of humor into the whole thing.

A popular song at the time was "Whatsa Matta You, Shut Uppa You Faze," so Don used this little song title as a substitute for "no statement found."

All would have been OK except, in our infinite stupidity, we managed to ship the system with the English version, and the good professor called me at home from Italy one Sunday and was furious. He said if we didn't send the Italian version *immediately* he would return the entire system and we could forget the order!

Panicked, I went down to the computer lab and rummaged around until I came across a data tape reel that said "Preliminary Italian version." I grabbed the tape and got it to an air express terminal. Away it went to the good professor.

About a week later I received the following FAX from our chief of operations in Europe:

TX NO: 771   30.9.81
ATTN. MJC
SUBJECT: NAPOLI MUSE STATEMENT LIBRARY
(ITALIAN)

THE NSF (NO STATEMENT FOUND) ACRONYM WAS TRANSLATED "WHATSA MATTA YOU, SHUT UPPA YOU FAZE."

I WOULD APPRECIATE YOUR IMPRESSING ON THE AUTHOR THAT THE DOCTORS INVOLVED WITH THE SYSTEM ARE A WELL EDUCATED AND A SERIOUS GROUP OF USERS AND WOULD APPRECIATE MORE RESPECT.

REGARDS: TONY WAKLING

THE IRON WILL TO LIVE

In 1988, Marquette acquired a company in Bangor, Northern Ireland by the name of Temtech Ltd. The company was one of the first to perfect and manufacture a device called the defibrillator, a device used to electrically shock a patient out of rhythm disorders of the heart that would otherwise be fatal.

One of man's greatest inventions in my view.

Just think about it. A man or a woman lies unconscious on the floor, soon to die from irregular rhythm in the heart, and along comes someone with this device, about the size of a large shoe box, puts electrodes on the chest, and within seconds the lucky soul is shocked back from the dead to the living.

Where else could you find such a simple device for saving lives? Every year, hundreds of thousands, perhaps millions are saved by defibrillators. At our Milwaukee plant we used to have an annual celebration where the mayor and the fire chief (fire trucks were equipped with the devices) would come with people whose lives had been saved by one of these amazing gadgets – along with the firemen who did the saving.

Up on the stage of our company restaurant they would present various awards and then we would have a chance to talk, face to face, with a few people who had actually come to the brink of death, yet were saved by a shock from one of our defibrillators.

I asked one fellow what it was like to be dead and all he could say was "quiet, very quiet."

Temtech was small – perhaps fifty people – yet the workers were very serious about the equipment they produced, recognizing that it was used to save lives. Although there was, as usual, political unrest with the North and the South, trouble that often materialized in violence next door in Belfast, the little Irish lasses on the assembly line worked side by side in complete harmony even though one might be Catholic and the next Protestant. The assembly process was slow compared to modern robotics methods, with each component selected and placed in the circuit boards by hand, yet wages were low so the cost of producing was tolerable.

The chief engineer of this nice little Northern Irish company that made these defibrillators was a tall, slender fellow by the name of Harry Magee, and a nicer, smarter, more kindly fellow you'd never want to meet. Harry would come over to Milwaukee quite often and Lisa and I would always have him for dinner. We'd talk 'til the wee hours in the morning about engineering, business, religion, philosophy, and any other subject that would come to mind. He became one of our closest friends from across the sea.

One day in late 1989, having returned from a business trip in Australia, Harry took ill, very ill, and no physician, no medical expert of any kind seemed to be able to determine Harry's problem, much less treat it. Within days he lost his ability to move his legs, his arms, his body – and soon Harry was totally paralyzed. He had lost all motor control except the ability to move and blink his right eye.

As near as we could tell, and the experts all agreed, some wicked virus had ravaged Harry's body from head to toe, leaving only his right eye and internal organs to function. For months he lay totally paralyzed in his hospital bed, unable to move or eat or speak, sustained only by intravenous feeding.

Finally his family convinced the doctors to let Harry come home and Joan, his wife and a nurse as well, would care for him with the help of the kids.

They did. He lived on and each day the prayers went out for some sign of recovery of at least a finger, a toe, the other eye – something that might be restored at least to some small degree of mobility, but almost no improvement was realized. Mentally, Harry was alert and trying, as only a tough, Irish soldier could try, to overcome in some way the plight that had besieged him.

We continued to pay him for several years and, after that, I offered some of my own company stock for the family to cash in each year.

Somehow, I felt our vast connections with the medical profession could be harnessed to bring help to Harry. That somewhere there must be a device, a treatment, something like the defibrillator where we could press the button and Harry would be well again, but there was nothing anyone could do.

He could communicate only by answering questions with the blink of his eye for yes, two blinks for a no. His wife agonizingly and patiently wrote letters for him by asking multiple choice questions, receiving "yes" or "no" from Harry's eye blinks. With his determination to, somehow, continue to do useful engineering work in spite of his condition, we got a computer for his wife to operate, using Harry's eye commands to guide her.

Soon he wrote:

17 January 1994

Dear Mike,

At last PC ANYWHERE is working, and I can't tell you what it means to be back in communication again. Once more I am indebted to you.

I have another request, which I shouldn't have the cheek to ask for, but I have. So here goes (at this point you're probably wondering "what the hell does he want now?") Has M.E.I. (Marquette) any circuit simulation software that I can install?

I would like to keep my hand in circuit design (I still know a bit about defibs and a little about electronics). With a graphics printer (which I have) I could run and test circuits on the simulator, then print them. As you may have gathered I am looking for something useful to do.

Harry

In spite of it all, Harry kept his sense of humor, a wonderful Irish trait. How could anyone keep their sense of humor after becoming totally paralyzed, pray tell?

Later, I told Harry I knew of a "headset" device used for paraplegics that could track movement of the eye, thus opening the possibility of his operating the computer by tracking his eye while looking at an alphanumeric display and then entering a command with the blink of his eye. We worked with several engineers on the idea, and finally installed a raft of equipment in the bedroom of Harry's home.

With all this Harry was able to *write* although the process was painfully slow. His will to succeed in this process is reflected with these letters I received:

1/17/96

Dear Mike,

Thanks for the shares (of stock I sent him) believe me they were invaluable.

I am using the headset but it isn't getting any easier. There doesn't seem to be much point going back to the old system. I have exhausted the possibilities in that. So I am committed in using the new system and that's how it should be. You know, better than anybody, that you have to keep moving on. So I persist. I don't know (yet) if PC Anywhere needs to be reinstalled for Windows or if it will work as it is. Until I find out I am sending this by air mail. Please let me know if you get this OK.

Best wishes to you and Lisa

Regards from Harry and Joan

I will be the first to admit that I had trouble in answering these letters. I found it was easy to get choked up and procrastinate. One side of me said there was no hope for Harry – the other that one day he would suddenly burst into recovery and I must do everything in my power to help.

In this next letter he speaks of a "blow-suck" switch that has a small tube and switch one puts in their mouth. By blowing or sucking, then you can say "yes" or "no," but Harry evidently had so little control he could not make it work.

May 26,1996

Dear Mike

I was suffering from a fundamental mistake regarding the use of the headset with an eye switch. When I move my head it causes an involuntary *blink* which I cannot control. I think the manufacturers realized the problem and provided a variety of different types of switches which would allow you to move the cursor to where you want before *blinking*. Note that the *blink* is intended to be a suck-blow, or a movement of the thumb, etc. Basically, there is a switch designed for whatever mechanical movement you can manage. Unfortunately, at present, I can manage none of them. The only kind of movement I am capable of is to blink my eye. Problem is, an eye switch is a no-no for the headset. I have thought about it long and hard and have come to the conclusion that until my condition improves to the point where I can use some of the available switches or somebody devises a way round the problem Windows is not for me. I have even considered a thought-control switch. (I got the idea from a paper you sent me).

I look forward to hearing from you soon.

Best regards,
Harry

I have never known, nor will ever know, I am sure, a human being with so much courage, so much fortitude to live on, in spite of all odds. Harry died peacefully in his sleep on a spring day of 1999, almost ten years after the dreaded virus attacked him and paralyzed every part of his body but his right eye.

May God bless Harry Magee.

ROCK KENNEY'S MARRIAGE

Somewhere in the late eighties we hired a wonderful, big, stout Irish-type guy by the name of *Rock* Kenney to drive our pickup van. He would go to the airport and fetch visiting customers or take one of us to O'Hare Field for an overseas flight or whatever errand that needed to be done. Rock was *The Rock* because he was so reliable, never failing to find some poor, lost customer on a delayed flight.

I wouldn't allow the company to buy long limousines because I thought they were ostentatious, and Rock agreed. It was part of our image to present a sort of down-home, unpretentious style to everyone because we were not trying to put on a show, just get the job done. I think our customers appreciated that.

Rock was very busy with this job, and soon we had to get another van and hire a second driver. Eventually, he had quite a staff. We never let visitors come into town without picking them up. We thought it was one of the many ways we could show courtesy to our guests, and I'm sure they appreciated that. What could be worse than to arrive in some strange town on a cold and blustery Wisconsin winter night and have to find a cab?

One day at our 1991 annual stock meeting, Rock stood up in front of the crowd of some two hundred people gathered in Le Bistro and announced that he was going to marry one of our best secretaries, Mary Hohensee. This was not a normal kind of announcement for our stockholder meeting, but then, we never did do things normally. We had lots of intra-company marriages in our organization. In fact, half of the day-care kids were *Marquette kid*s, so it was said.

It took three years after his dramatic announcement for Rock to raise enough money to actually tie the knot. But in the summer of 1994 he was ready, and an old buddy of mine working in our PR department, Bill Browne, said, "I have a really crazy idea for Rock's wedding that's coming up next week, Mike. Want to hear it? I'll warn you, it's pretty far out."

"Sure, Bill, let's have it. You know me – the farther out the better."

"Well," says Bill, "why don't you rent a chauffeur's outfit – you know, cap and all, and then you pick up Mary and Rock on their big wedding day and then drive 'em to the church?"

"WHAT?" I said, somewhat stunned.

"Sure," Bill said, "and I know a place where we could rent a 1975 Rolls Royce to really top it off!"

I had to admit this was one of the greatest ideas I had ever heard. So, on a beautiful Saturday afternoon, I showed up at Mary's house and opened the back door of a big, white Rolls sedan and said, "Good afternoon, miss. I'm here to get you to the church."

Mary sputtered and said, "You look just like our president, Mike Cudahy. In fact you *are* Mike Cudahy, aren't you?"

My next mission, of course, was to pick up the groom, and this time the newspapers were tipped off. I again jumped out of the Rolls, opened the back door and said, "Good afternoon, sir, I've come to fetch you for your wedding."

Rock didn't get it. A news photographer snapped several pictures, making sure I was bowing and politely getting Rock into the car, but Rock just had a bewildered look on his face.

Finally, as we drove away, he leaned forward in his seat and said, "I don't know how to say this, driver, but you look just like my boss, Mike Cudahy, and that's why I'm acting sort of strange."

I couldn't stand the suspense any longer and said, "You mean that creep that runs Marquette Electronics? Listen, mister, I wouldn't want to look like him for all the tea in China," and the

Rock Kenney, chauffeured to his wedding by Mike Cudahy

jig was up. Rock burst into such hysterics I really didn't know if he could regain his composure in time for his wedding.

These stories represent at least some of the Marquette culture although, perhaps, in an indirect way. As one can tell from them, Marquette had a reputation of being a very fun, a very gratifying place to work.

INC. magazine had a lot of fun in 1985 with a feature article on Marquette, rating us one of the best managed companies in America. They took a shot at explaining our culture but went a bit off the deep end with their opening remarks as follows:

WILL THE COMPANY PLEASE COME TO ORDER

Who's in charge at Marquette? (Hint: It only looks like it's the employees).

Quite frankly, the executives at Marquette Electronics, Inc. would be a little uncomfortable with the notion that theirs is one of America's best-managed companies. Here's a manufacturer of sophisticated medical devices used by doctors around the world to make life-and-death decisions. And yet, here's a company organized around the unorthodox belief that there is productive value in fun and a creative merit in a bit of disorder. By any standard, it is an odd combination – as a group of visiting West German businessmen discovered.

As the story goes, the foreigners were taking a mid-shift tour of Marquette when they happened upon Hawaiian-shirted employees with boom box hula music playing in the background. Puzzled, the visitors turned to their guide, Michael Cudahy, Marquette's president.

"Just having a little fun on their coffee break," Cudahy explained.

In spite of such bizarre writings, Marquette could well be described as a different kind of place to do business with as well as to work for. Explaining these subjective qualities, however, is not easy. True, those who have worked at Marquette and those who have been good customers know this very special, abstract quality that we, perhaps by accident, created.

Visitors used to say they could feel it the minute they walked in the door. Then I'd ask them to describe the feeling and they couldn't. Perhaps they shouldn't. Perhaps it's one of those things better left unsaid.

I asked one of my close compadres, Kevin Lindsey, and he said the feeling when he joined Marquette was that finally someone trusted him. That no one was watching over his shoulder. That if he wanted to try a new idea, he was sure no one would crucify him. He said he felt, for the first time in his working career, that the sky was the limit, and if he stumbled, so what? Try again.

CHAPTER 10

FAIR PLAY

In the previous chapter, I hope I've given you an idea of the basic fabric that made up our culture – things like Le Bistro, and how we served beer and wine. Of the parties when we were successful with some new product. Of the day-care center with all of those kids and the fuss we made over their graduation every year. Of how we were one big, happy family, even when there were thousands of us.

Also very important was how our employees were treated, financially, and what the physical surroundings of the workplace were. And how our profit sharing and ownership plans worked for everyone from the janitor to the CEO. And, yes, how we treated our customers and suppliers.

When I started Marquette in 1964, I had only one employee, Dan Phillips. I reasoned that if I was fair with him he would probably stick around and work hard. He was very valuable to me so I decided I'd do my best to make him stay.

More than being fair, I came to the conclusion that if Dan liked his job, he'd probably do better work than if he didn't. This simple idea grew to become the framework of how we treated all employees through the years. Try to see to it that people are having a good time – perhaps even loving their jobs.

"Spoil your employees," some would say, "and they'll take advantage of you, sure as hell!"

On the contrary, my experience tells me that most of our work force was honest, sincere and liked to work hard. If treated fairly and given opportunities, they would bust their gut for the company, as long as they were enjoying their job.

At the start of the industrial revolution, it seems that man somehow got the idea that discipline and scrutiny were the only ways to control the so called *worker* – that keeping a careful eye and a tight leash on him would improve production and lower costs. The idea was that the company would be more profitable if pay and expenses of the worker could be kept at a minimum – and to hell with how he liked his work.

Not surprisingly, this attitude was, in my view, the seed that started the modern day labor union. Suspicion (on both sides), unfair pay and miserable, if not unsafe, working conditions caused workers to do what was only natural for them to do. That is, gang up and form unions which could possess enough power to fight back and combat management abuses.

Once this kind of mind-set got started, it's not hard to understand all the ugly adversities that grew from it. Polarization between management and the work force created bitter bargaining sessions for wages and working conditions rather than cooperative planning with mutual trust and teamwork, fairness on both sides in the name of greater production and profits, job security and eventually even greater pay.

Marquette pursued the idea of fair play with vigor. We started with a clean piece of paper. We decided that we would give everyone a nice, safe, and comfortable place to work. Instead of time clocks, we told everyone we'd trust them to put in a good day's work. And we gave (and sometimes sold) the gang a piece of the action to assure them that they could call the place, at least partially, their own.

At first, we offered employees company stock for cash or pay deduction. No lofty contracts. I'd just say, "You want stock in the company? OK, I'll sell you some, and at a fair price." The "fair price" calculation was halfway from our book value to what we all thought would be the share price if we had been a public company at the time – in other words, sheer guesswork.

Guesswork or not, a lot of employees who stuck around made a pot full of money. Sometimes we gave stock away to say "you've done a hell of a job." Then there were options – lots of options.

As far as the paycheck was concerned, we researched what was considered the going wage for various work skills at other companies and freely shared this information with everyone in the company.

I never did subscribe to the simple-minded notion that paying more money made for a happier employee. Sure, there's clothing to buy, rent to pay, and mouths to feed. But too many employers think that more money will fix the problem without even understanding what's going on in the heart and soul of a disgruntled employee.

More important to us was to make a concerted effort to see that everyone had a job not only that we thought they could do but one that the employee *wanted* to do. Often, we had to make adjustments to accomplish this and often it required training – schooling on the job. The decision to operate in this manner was a very conscious business decision, not a do-gooder quirk of mine. We all felt that this sort of way to treat people was the best way to get the most out of everyone and the best performance out of the company.

It was. In spades, it was. And we never had that nasty "you sit on that side of the table, and we'll sit on this side while we wrestle our problems to the ground." We never wrestled. We never attracted a union because no one wanted one. Everyone felt they were being treated fairly, and they were.

In the military, I learned (the hard way) that you have to do what you're told – or else. That's the way it is, and perhaps that's the only way it can work in the service. Do what you're told or they'll give you a worse job. If you won't do that, you get court-martialed.

In civilian life, things are obviously quite different because in a free society, there are alternatives. If you don't like the job you're doing, you won't put much into it or you can simply walk out and find another. This is particularly true when unemployment is low and jobs are plentiful. Put another way, the onus is on the employer to keep the employee, much as some employers might not like the sound of that. We came to the brilliant conclusion that everyone had faults, by the way. The trick was to work around them.

Our attitude of bending over backwards in these various ways created a work force that had an unusually powerful drive to push themselves beyond their capabilities. And it developed an extremely strong will to stick around. Turnover, the most expensive problem in any industry, was very low.

Adding to the idea of selling stock and giving options for a piece of the action, we implemented a rather conventional employee stock ownership trust (ESOT) in 1974, transferring a percentage of our profits to everyone from the janitor on up in the form of stock. The board figured the amount each year, sometimes reaching as high as 12 percent of our profits. In 1988, we also started a 401(k) plan where employees would put a part of their pay into a savings account for retirement and the company would sweeten the amount with another 25 cents for every dollar.

The physical surroundings at Marquette were quite different as well. I always felt that a factory didn't have to be so cold and unfriendly as the many other establishments I had seen through

Monitor final assembly and test

the years. Again, I took ideas from Harry Bradley of the Allen-Bradley Company that the environment could just as well be friendly and warm.

Most companies call in a professional decorator to make sure things are *color coordinated* to look attractive, at least in the front offices. Never mind the personal taste of the people who live in these places. The policy usually applies to everything except the executive offices. Here the decorator has extra money to spend for *personalization*, yet there's still too much uniformity. Wall colors, pictures, furniture, even the shape and size of offices are uniform, so that space can be used most efficiently. The factory, of course, ends up being neglected – gray and beige with only the drone of a paging system for a personality.

With all this conformity, a lot of work places look like chicken coops, and the occupants tend to feel like chickens. Besides that, it's demeaning as hell to have to count out the aisles and rows to

JOYWORKS

make sure you don't sit down in the wrong office when you return from getting coffee. Even worse, to be told that your kid's picture must be desktop sized, and you can't keep that ivy plant you brought from home because things like that are provided only by the rent-a-plant company.

At Marquette we pretty much let everyone build their own corner of the world. We gave them almost unlimited freedom. Employee attitude, as I keep saying, is one of the most important concerns in business. So if this helps, why not? Uniformity for the sake of uniformity, rules for the sake of rules, well meaning as they may seem, can end up costing the company many, many hidden dollars simply by creating poor morale.

Dress code? Why not let people come to work in any kind of clothing they like. I used to come to work in shorts on the hotter summer days. Once, a new employee asked one of our old-timers who that man in shorts was walking down the hall, and the guy told him that was the boss.

"You mean the CEO comes to work like that?"

"Yep, young fella, and I can see you're going to have some getting used to in his outfit," said the old-timer.

Sure, when some distinguished professor would come to visit us from overseas, we all knew that a coat and tie was the order of the day. We seemed to know what was right and what was wrong on any given occasion. But where is it written that a dress code improves the bottom line or gets the order sooner or improves the product quality? Maybe that's something Henry Ford said at one point in his passion to turn people into monkeys.

In 1983, I wrote a memo to the troops bufooning the whole idea of a dress code while making sure everyone knew that when we had a visitor of great importance and dignity, it was a good idea to dress appropriately:

132

To: All Personnel
From: Michael J. Cudahy
Subject: Dress Code

Summer is coming, and with it, hopefully, warm weather, so here's a word to the wise.

There's no dress code at Marquette and people, I suspect, often wonder what that really means. No coat? No tie? Shorts? Jeans? Bare feet?

Well, first it means common sense. Bare feet are somewhat dangerous around the machine shop. Roller skates might be OK if you don't go too fast and crash into someone. A few blobs of acid on your knees might give you the message that shorts aren't too smart in the circuit board fab lab.

I really only have one problem as to how you look, and that's related to the never ending plant tours by customers. If you're a so-called *boss*, it's not too smart to show up in tennis shoes, shorts and no shirt if the Chief of Cardiology from the University of Madrid is expected. Playing *boss* with a three-piece pinstripe suit also looks a little stupid if you're not one. Jeans make it hard for people to know what sex you are, but maybe you don't care. If you're a receptionist or tour director, it's obviously a good idea to look sharp and neat.

But, as I've said many times, we're on the cutting edge of a new industrial revolution with our company, so when we aren't going to have visitors, wear anything within reason – but at least wear something so you don't get arrested on your way to work.

A year or so after I sold the company to GE, they released a memo to the troops, dealing with dress code, as follows:

To: All US GEMS Employees
From: GEMS Leadership Team
Date: April 11, 2001
Re: Business Casual Dress Policy

Every interaction we have with our customers and visitors at our facility leaves a lasting impression. And this year, we expect to have more visitors than ever before. In fact, every month we have as many as 100 doctors, nurses and hospital administrators, and more than 30 employment candidates visit GEMS facilities.

Therefore, as we invest in renovating our facilities to present a more professional, technology-oriented image to our visitors, we will also be adopting a business casual dress policy. Specifically, this means that sweat pants, shorts, t-shirts and tennis shoes will no longer be acceptable attire in nonproduction areas starting May 1st. At the same time, blue jeans are discouraged for anyone who routinely comes in contact with customers or visitors.

As a leadership team, we have talked with many people while considering the change. The feedback almost universally is that a formal guideline is needed to assure we project the highest standards of professionalism that surpasses our competition and represents the industry-leading values of GEMS.

We hope that you agree that every interaction with our customers and visitors can make a difference. Thank you for your support.

This memo, of course, leaves no room for common sense on the part of the employee. Just follow the rules. Without rules how can a company like the one we used to run have the *professionalism* GE speaks of?

Answer – with a zero dress code, with no time clocks, we, somehow, had the highest reputation in our business. We were not liked, we were loved and respected by literally every customer we had. And we had the most spirited, hard working, creative work force in medical electronics. Our customers could feel it in the air. We beat every competitor we came up against, including such names as HP and Siemens Medical.

"Wait a damn minute," you shout. "In spite of what you say, how did you prevent people from running wild. Why didn't they go crazy and why didn't the place look like a zoo? Sloppiness must have cost the company an arm and a leg."

"Not true," I say. Most people control themselves, and Marquette was living proof. The troops wore appropriate clothing and arranged their surroundings for the best way to get the job done. True, they made *work nests* by putting their personal affects around them and on the walls, but they were reasonable about it, and that was just what I wanted to see happen. Their *nests* became an extension of their home and this, believe it or not, helped keep attendance high. In fact, we often had people come in on weekends when the home front got uncomfortable.

Outside, the only reserved parking places at any of our plants were for visitors. I can't imagine what people are thinking when they make a nice little space for the CEO, the President and a VP or two right at the front door.

How would you feel as a worker on one of those rainy, sleeting days when the only space to park is way in the back lot? Now, as you walk towards the plant, freezing your tail off, you spot the CEO sliding into his special spot at the front door.

Wouldn't you think up some nice names to call him under your breath?

Inside and out, Marquette always looked neat and orderly to the outsider, and the reason was that everyone was proud of every inch of the place because they owned a piece of it.

And we always kept management out of the personal problems of the work force. If someone came in to complain about a fellow employee playing a radio too loud, I'd say, "That's your problem. You and the others will have to settle it among yourselves." I felt that if management got caught up in that sort of thing, there'd be a never ending battle and management would lose. If management decided to *control* the loudness of radios, for example, we'd have to get sound meters and hire a staff of snoopers to check every boom box in the place.

One day, three young gals from the back of the plant came to my office to tell me of a big problem they had. (I always had my door open so if anyone had a problem, they could just walk in.) They said, "Mike, you've got to do something about those guys back there in shipping! They've put big pictures on the walls of almost-nude girls, and we think it's disgusting!"

"Your problem," I said. "Management can't get mixed up with personal problems in our outfit. You'll just have to settle this thing with the guys back there on your own."

They left a bit miffed, I might say, but reluctantly accepted my answer. I could tell, however, that there was going to be a battle. About a week later, I found out what they were up to. The gals went out and bought two huge posters of muscular, macho guys, almost nude, and tacked them up in a prominent spot right near shipping.

By the way, there were some restrictions we all had to adhere to. You couldn't smoke where they were putting together delicate electronic equipment because the smoke raises hell with it. And we had to tell one shop lady not to bring her gerbil cage to work anymore because we were afraid the FDA would see it on some inspection tour and *write us up*.

These are, perhaps, good examples of our business style. But as I write these words, I still don't seem to identify the warmth, the magic, the brotherhood that seemed to exude from the walls. What was it, for example, that made our customers feel like they, too, were part of the family? Perhaps it was that our customers were.

Way back in the very beginnings of Marquette, I started making friends with the chiefs of cardiology at the major institutions in North America. It happened quite naturally because the products we were selling were expensive and in those days the chief was whom you had to sell – not the administrator, as was the case when managed care came in. Nor the president of the hospital, although that acquaintance sometimes helped.

Basically, it was the chief of cardiology who made the decision on our equipment. What he said was what usually happened, hang the cost.

These chiefs were really outstanding people for the most part and the kind you'd like to have as friends. I never schmoozed them, as the old saying goes – the thought never occurred to me. I just got to know them because they usually wanted to get to know me and the new machine we had developed for Northwestern Medical School. What I didn't realize was that I was developing lifelong friendships with some of the most interesting people in the world.

So many friendships in business are phony, in my view. "You'd better like old Joe Jones because he's the guy who signs the order" kind of thing. I'd have to say that the pals I cultivated in the business of medicine were really nice people. I never felt *put upon*, nor did they.

When I first started dealing with the Mayo Clinic, for example, I knew the chief of electrocardiology, a terrific guy named Ralph Smith. He became a lifelong friend. In knowing him, I got to know another dozen or so great cardiologists at Mayo.

And in those days Mayo used to let a large number of the doctors go to various annual meetings such as those for the American Heart Association and American College of Cardiology. On the first night at those affairs I'd always round up the Mayo gang, and we'd all go out for dinner. When the check would come one or the other would say, "Well Cudahy, I suppose you're going to pull your usual line, 'We'd better make this Dutch treat fellows, because you wouldn't want to be in a position of being influenced by a vendor, would you?'"

Accordingly, we'd all chip in our share with a pile of money, just like a bunch of old ladies out for lunch. We'd divvy it up and argue like cats and dogs about which drink was which and who was the guy who had the expensive desert. There was a great, warm feeling of camaraderie that went on year after year, and no one ever felt that he was being *bought* or influenced about making a buying decision.

Customer-friends also had a lot to do with our innovations. We spent much time at our favorite hospitals, snooping around to see what their problems were and how we might apply technology to help. We found if we listened in a truly receptive way, we'd often come up with something that would be useful not only to them, but to the rest of the world as well.

I had such a customer-friend from the Medical College of Wisconsin, who was famous for kidding around in a gruff sort of way, but a wonderful source of new ideas. John Kampine, MD,

PhD, and Chief of Anesthesia at the college was about six-four. Very big, very imposing and very brilliant.

He called me up one day somewhere in the eighties and said, "Come out here to the medical school. I have something I want to talk to you about."

"OK," I said, mostly because I loved the guy, and it was always fun to go to his labs and get an education, no matter what he had to say.

"You see that scope over there with the lousy signal trace on it?"

"Sure, John, what are you trying to see on the damn thing?"

"Ischemia. When I put a patient to sleep, it's a very good idea to see what his S-T segment is doing before it's too late. And on that stinking little scope – I'm supposed to judge if the guy is going south on me or not?"

"Yeah, so?" I said.

"Listen, stupid, your company has been measuring S-T segments on people running on treadmills forever, and you don't seem to have the sense to adapt that technology to the operating room."

"Hell, John, I could throw something like that together in a few weeks if that's all you want. No big deal."

"Well, then do it, damn it, so I don't get eye strain around here trying not to kill my patients."

A month later we had a prototype, and shortly after that our operating room monitors had this great feature, and I told John I was glad I thought of it. A year later, most competitive monitors on the market had the same feature in one form or another, and one company even had the audacity to claim they invented it.

Let me present now five examples of business practices which took place at Marquette that were, in the conventional sense, totally off the wall. Were we nuts, or should some of the old ways be revised?

RESTRAINT OF GREED

Anyone intensely interested in the success of a manufacturing enterprise will tell you that, above all, you must do anything and everything to improve the almighty bottom line. Higher prices, cheaper cost of goods, lower labor costs – and above all, don't tell the customer anything about such inner workings.

Fundamental wisdom? Not necessarily.

In 1979, we landed one of the biggest contracts in our history with the U.S. Department of Defense (DOD). It started out by responding to a bid for "CAPOC" – a Central EKG System for all military hospitals, Army, Navy, and Air Force.

Needless to say, we worked very hard to present our company as being the most qualified, and this wasn't easy against such giants as Hewlett-Packard. But, after all, we were the first to introduce such equipment to the world and had installed systems of this type in most of the major hospitals in North America.

The $24 million contract requirements were a struggle for the first year. In fact, sometimes I wondered why we ever thought we could tackle such an undertaking. There were some 172 hospitals involved, and we not only had to build central EKG systems, but the bedside data acquisition carts as well. The years that followed, however, were paradise. We ended up on all sorts of Army and Navy part numbers so when a "GI" would need an EKG machine, he'd look up "type number xxx" and, automatically, order a Marquette product. The original contract turned into a virtual gold mine.

This all took place in the days of intense, out of control inflation, so the Department of Defense allowed suppliers like

Marquette to automatically raise prices by 10 percent each year for quite a number of years.

One year, when we were about to take another 10 percent goodie, I asked our accounting department how we were doing on the DOD contract. They said, "Excellent, Mike, we're cleaning up! Best account we ever had."

At that point, I suggested something very strange. I said "What if we decline our 10 percent increase this year? What if we write DOD and say we are nicely profitable with the contract as it is and don't need the increase?"

The looks around the room were that of utter disbelief. Cudahy has lost his marbles. He's flipped!

But then someone said, "We'd sure be a new kind of hero down in Washington if we did that, wouldn't we? And that could possibly lead to more business, you know." Another chimed in, "You know they might audit our books one day and say 'excessive profits' or something like that if we took the increase. Maybe Mike's right. Maybe now is the time to be good guys. After all, we're all U.S. taxpayers."

And so we declined our 10 percent increase with the U.S. Department of Defense – probably the only company in the history of the United States. DOD went wild. From that day forward, we could do no wrong. Our technical problems (and there were a few) were all forgiven. And if it was ever in their power to swing more orders our way, they did.

COMPETITORS WELCOME

Every year our company, like all other companies in medical electronics, attended various national meetings, conventions if you will, where papers were presented on the newest medical discoveries and manufacturers could display their wares.

This was of great expense to companies such as ours. Not only did we all pay dearly for the so called *booth space*, but we had to get there with all of our equipment and personnel and then set it up and make sure everything worked properly.

And then there was the sales and marketing force. Expensive people going to expensive places like Anaheim, California, Dallas, Texas, and Atlanta, Georgia. These characters had expensive tastes, too, when it came to going out on the town. But we had no choice in the matter. If we didn't show up, our competition would "have our lunch," or so we thought.

Every year I'd address the troops before the meeting and say, "Now, don't be a wise guy around the competition. Don't tell them how good business is – tell them it's pretty bad. Why? Well, dummy, if you say is great, they'll go back home and try harder. But if you say it stinks, they'll have some laughs, saying 'not to worry about Marquette. We've got 'em on the run.'"

But then the show would open and the competition would start the damnedest war dance you'd ever want to see. They would get phony badges saying something like Dr. Smith from XYZ Hospital and come in our booth asking all sorts of questions about our new equipment. Of course, 90 percent of the time someone in our outfit would recognize the guy and say, "Hi, Joe. What are you doing wearing Dr. Smith's badge?" Usually, with a red face, he'd have to confess he was sent on a spying mission.

So, with all this *I Spy* stuff going around, we decided we'd have a special "competitors morning" on one day of the show and invite all the competition into our booth, showing them everything.

Why not? Maybe we were a bit smug, what with leading the pack in new innovations, but we felt they could do little harm by seeing our new things first hand. Their reaction, usually, was to sort of pooh-pooh everything, having themselves a good dose of the *not-invented-here* syndrome.

OPENING THE BOOKS

I have always felt that there's not a single thing about making a profit that's shameful, as long as it's reasonable. Part of free enterprise, right?

"Of course," you say, but then why do so many people in business go to such pains to hide their profits? So the competition doesn't find out? Nonsense!

Let me give you a wonderful example of what I speak.

Sometime in the eighties, the president of Latter-Day Saints Hospital in Salt Lake City called me and said, "Are you Mr. Cudahy?"

"Yes, who's this?" I replied.

"Look, I'm Frank Kilborne, Chief Operating Officer of Latter-Day Saints Hospital in Salt Lake City, and you've got me in a heck of a spot!"

"Oh? What did I do wrong, Mr. Kilborne? Help me understand," I said.

"Listen, Mr. Cudahy, we're trying to purchase a whole bunch of monitoring equipment for our hospital and we always go out on bid when this many dollars are involved."

"I've heard you're going to buy several million dollars worth of gear, sir, and, needless to say, I hope you'll choose Marquette. But what's the problem? Can't we just bid it and hope we'll be the winner?" I asked.

"That's just it, Mr. Cudahy. You see, our doctors are so nuts about your equipment, they won't let me buy anything else, so how can I ask for bids? I mean, what if someone underbids you? I'd have a riot on my hands. You've got me over a barrel. You could charge me anything you wanted, and I'd have to go along with you."

"Tell you what I'll do, Mr. Kilborne. I'll come out there and bring our books. I mean our calculations on the profit for your equipment if we get the order, OK? Our accountants are very well respected so I guess you can believe the numbers."

"Mr. Cudahy, that would be wonderful, but I've never heard of anything like this," our prospective buyer said.

The next day I was on my way, loaded with all the calculations on Latter-Day Saints.

When I arrived, Mr. Kilborne, the Chief Operating Officer, was waiting and soon I had the papers spread out on his desk and explained the numbers. I said, "I assume you don't mind if we make a profit, do you?"

"Of course not," he smiled. You'd not stay in business very long if you didn't make some money."

"How much do you think we should make, Mr. Kilborne?" I asked.

"Well, let's see now. I'm not experienced in manufacturing profits, but I would think something in the order of 25 percent before taxes would not be too much. What do you think?"

"God bless you! Would I love to make a deal like that. Actually, we were hoping for a 15 percent profit before tax on your equipment if nothing went wrong – I mean installation problems and things like that, Mr. Kilborne."

"Where do I sign?" he said with a broad grin.

RESPECTING SUPPLIERS

We used to build our own *power supplies* for our patient monitors. That's a box inside the unit that takes electricity coming in from the power line and converts it to what the monitor needs.

A lot of small companies (and some large) make power supplies for manufacturers like Marquette. It's a sort of specialty business that allowed us to concentrate on more complex designs peculiar to monitoring.

So one day we decided to buy rather than build these little boxes. And the company we decided to buy them from was a little outfit in Northern Wisconsin run by a nice guy by the name of Bob Mast.

Things were going along pretty smoothly with this arrangement, and one day I asked our plant manager how much we were paying Bob for the boxes.

"$630, Mike," the plant guy said. "I'll be damned if I can see how he can make money at that rate, but that's the price we negotiated with him."

"Did anyone ever ask the poor soul if he might be losing his shirt at that rate?" I asked.

"Well, no, Mike. We were a little scared to ask him. You know he's our only source now, and he might just jack up the price fifty or a hundred dollars."

I suppose I was micromanaging again at that point, but I picked up the phone and called Bob Mast and asked him, straight out, if he was making or losing money on our account. He said he was, indeed, losing his shirt.

"Bob, I think you should raise your price to us by $100. That way you'll be around next year. You know you are our only supplier, and I want you to be healthy, not broke and closing up shop."

Bob Mast almost fainted! Pennies from heaven. "Mike, I'll make you the best power supplies you've ever seen. I'll be indebted to you forever," he said.

Now, was that good business judgment or bad? I know as well as anyone that increased prices from suppliers will lower our profits. Or will they? If Bob Mast went out of business, we'd have to find another supplier, and very likely we'd end up paying more.

Besides, he was a nice guy.

THE RED CARPET

Finally, there was the jet airplane we had, which one *might* say was influential in some buying decisions, speaking of customer relations.

It all started when I looked into how many potential customers ended up buying Marquette equipment after visiting our plant in Milwaukee. I found out that once these medical groups got a chance to see the operation and meet some of our engineers, production people and sales managers, they'd be so impressed that we'd usually end up with an order.

The problem was that it wasn't always so easy to get the *higher-ups* at the big institutions to just drop everything they were doing for a trip to Milwaukee. Chief of Cardiology Brown or Administrator Smith would say, "I'd love to see your operation, but there's just no way that we could spare the time to get out there in the near future."

One day it dawned on me that if we owned a jet, we could offer to pick these people up, fly them to Milwaukee and have them home, usually on the same day. That would be an offer they'd find hard to refuse.

Then I found out how much private jets cost. Good grief! $3 million and up. A nice, roomy one for $7 million. When I suggested the idea to my board of directors, they thought I had gone off my rocker. "You want to do WHAT?"

A few sleepless nights later, I decided I'd just buy a jet out of my own pocket. "What the hell, Cudahy," I said to myself, "you can borrow the money from the bank and then charge Marquette for the use of the plane. If it brings in business no one will argue, and if it doesn't – well, you'll just have bought yourself a very expensive toy you can't afford. Now you'll have to sell it."

But the idea worked. Over and over Dr. Brown and Administrator Smith would say, "Private jet? Sure, I think I can get the gang together. How about next week?"

Our pilots, Bill Lothman and Tommy Miller, who were great salesmen for the company in their own way, would go down the night before so the medical group could get going the first thing the next morning. They'd be at our plant in Milwaukee in time for coffee, a big screen slide show about our history, and a tour. Then down to business until mid or late afternoon. They'd usually be back home in time for dinner – of course bubbling over with enthusiasm about the trip. The success rate in getting the order with this routine was tabulated at 83 percent. Our competition was flabbergasted.

This was all part of Marquette's business practices. Later, when GE bought the company, I offered to let them use my jet on a sort of *time-share* basis so they, too, could reap the benefits of the magic we had created. They said "sure" but never used the plane for customers; rather, they flew their executives around the country.

Michael J. Cudahy and Warren B. Cozzens, founders of Marquette Electronics

CHAPTER 11

CHANGING OF THE GUARD

S omewhere in the nineties, I revisited the notion that all this talk about what would happen if I got hit by a truck was more than a myth. That one day I might be gone and leave some sort of unmanageable mess behind at the expense of my wonderful friends and employees. I wasn't really getting paranoid about trucks, but I was now realizing that I wasn't going to live forever. And that if I croaked, Marquette wasn't going to go on forever in the same way that it had.

Could my kids take over? I had five (with various wives), two boys and three girls. The two youngest, Mary and Mike Jr., were too young for such a responsibility. The oldest, Susie, was happily married to a FED-X pilot, John Johnson, and my oldest son, Patrick, is a commercial photographer. Julie, the gal who started the day-care center, didn't want to get into Marquette that deeply.

But what about Warren Cozzens' son, Todd?

This giant six-foot-seven, hawk-eyed fellow is a figure to behold. With IQ of equal size, Todd might have been the man to drive our company to the size of General Motors – or cause it to self-destruct.

Todd had been working for Marquette in many capacities when, in the mid-nineties, we assigned him the awesome task of revitalizing our operations in all of Europe. He wanted to take on Asia at the same time, but his father would have none of it.

Todd Cozzens lecturing to Europeans

Todd used to stay with us at our home called "Hilltop." He was always on his way from here to there, catching an early plane after staying up way too late the night before. From his room I'd hear, "Oh God, I'll never make it."

Out the door with a crash, his room would look as if a bomb had exploded in the closet. On the driveway I'd find all sorts of Todd's personal items including socks and maybe a shoe or two. His open suitcase probably spewed personal items all the way to his destination.

Europe will never be the same. Todd went from one country to the other, learning every language (including Russian), meeting doctors and making sales. No one would dispute that Todd could get the order – it was just a question of *on what terms*. There still may be doctors in Europe waiting for some feature of our equipment that Todd had promised – but never existed.

Could Todd possibly be the guy who could become number one at Marquette? Could he be stabilized sufficiently to run a major medical electronics company?

Probably not. He was like a red Ferrari, but all of us agreed, including his father, this Ferrari was a bit too fast.

I was particularly sensitive about the subject of succession because my old man was a good friend of Harry Bradley. Harry and his brother founded Allen-Bradley of Milwaukee, the world-famous electric motor control manufacturing company. Starting back at the turn of the 20th Century, they figured how to apply *carbon-pile* technology to motor controls so as to make it possible to run things like elevators. A truly great innovation.

When I was about sixteen, Bradley used to sit out on my old man's screened porch and rap about business. They'd let me stand nearby and listen, and I'd try to be cool and understand what they were talking about.

"I tell you, John, I'll never let anyone take over the Allen-Bradley Company, and I'll never let it go public either, do you hear me! Why, I have trusts written by the finest lawyers in New York City and five trustees to carry out my orders if I ever die."

"Now, Harry, just keep calm or you'll have a heart attack," the old man would say. "Michael, go get Mr. Bradley another drink."

"Yes sir, Mr. Bradley," I'd reply, and even at the tender age of sixteen, I felt I was in the middle of such an important meeting that if the drink I made wasn't right, it might cause of the end of the world.

In spite of this conversation, however, Harry died at the ripe old age of eighty-five and all but one of his trustees betrayed him. The company was sold for $1.65 billion. Harry's still spinning in his grave.

The point, and it's a very important point, I learned from that experience, is that no one can *freeze* time relative to the status of any company no matter whom or what or when. A company, I concluded much later in life, is like a living organism. It must grow and change shape or size, or both, or it will die.

So trucks mowing me down got to be a thing going on in my head. I was determined not to make the same mistake as old Harry – trying to freeze time.

And when my friend, Fred Luber, started badgering me again about finding myself a new chief, I was beginning to think seriously about it. I was keenly aware I should be looking for a guy who would be my sort of understudy for awhile and could eventually step in when I declared myself senile.

Fred tried to find his own replacement at his very successful fabricating company, Super Steel Products, and totally failed, so why not listen to what he did wrong. I also remembered the fiasco when I tried to get a guy by the name of Rick Balda from Hewlett-Packard to be my number two guy back in 1980. But just because I failed and just because Freddie had failed, well, that shouldn't be my excuse not to try again.

First thing, I reasoned, is this understudy can't be a clone – a look-alike in any way. If he is, he's doomed before he starts because the employees immediately make a one-for-one comparison. Since the incumbent is the *good old shoe,* any new guy is an ugly impostor or alien, and he's dead in the water on the first day.

Second, how to get one of those *executive* types to even vaguely understand our screwball outfit, which grew up from nowhere with crazy inventions that just happened to catch on? That would be one hell of a challenge, now wouldn't it?

I needed somebody who had reasonable technical and business skills but was a real human being. I didn't need the kind of skills one studies at Harvard Business School. I needed a person with the kind of personality that was cultivated by good fellowship – in going out to dinner with the gang and getting a little loaded together – in telling jokes – in ribbing each other about dumb judgment calls in business.

"Remember that contract with the Jones Company you signed, Charlie? How stupid can you get?"

"Yeah, Mike, but how about that time you told that customer from Cleveland you'd give him all new software for free?"

That sort of thing.

Not that that's all of what makes the world go round or that *professional* skills aren't important and needed. It's just that so often fellowship serves to weave the fabric of business together, and without that, all the professional skills won't amount to a row of beans. I don't think really true friendships can be crystallized without a little fun and games, and I don't think business can function without friendships.

Another thing I needed was someone who would take action. Have the guts to defy me or anyone else. I don't know if Freddie Crawford, the guy who started TRW, kidded around like that. But Crawford wrote a little booklet on business practices and in one part of this booklet, he talked about making decisions. He said, "Take action – move forward – even if you're wrong."

I agree. The CEO of any company must take action on many occasions, make decisions, even if he doesn't have all the facts and even if he could be proven wrong. A lot of times it's just plain dead reckoning. It takes a lot of guts at times to do so and face the consequences of making a complete fool of yourself.

I know. I've been wrong many times. Fortunately, I have been right more than wrong, but only by a slim margin. And I realize it was easier for me to take chances than anyone else. I was the founder, major stockholder, and always regarded as the boss.

The notion that someone could come in from the outside and make important decisions as the new boss is far-fetched, indeed. The only one who could succeed in that sort of environment would be someone with a ton of balls – operating in almost reckless abandon for his job or for the future of the company.

A big, red Ferrari gone wild, but Warren's son, Todd, was a bit *too* wild.

With all this in mind, I got on the phone one day to Fred Luber asking if we shouldn't have one of our lunches where we catch up on things.

"Sure" he said, "Pick me up in front at noon, OK?"

"See you, Fred," I said, and hung up.

At lunch, he threw a new idea at me to solve this "get hit by a truck" thing. "What if you could get the sharpest, youngest, rising business star in all of Southeastern Wisconsin to step in as president of Marquette, Mike? I know a guy who's just that. Barry Allen. He's now the president of the telephone company and an absolute whiz kid. Why, he's about to become president of Illinois Bell!"

"That's why you shouldn't drink at lunch time, Luber. You get goofy as hell, with just one lousy Early Times. What, may I ask, does a utility president have to do with Marquette Electronics, if I may be so bold?"

"This guy is different, Michael. This fellow isn't your usual utility bureaucrat. This guy is quite an admirer of yours, and he wants to be an entrepreneur just like you. Why, I do not know, but he does."

"Got to be nuts, Freddie, but so what? You just don't press a button on your head and, bingo, you're an entrepreneur, you know. The guy's probably an MBA or something really awful," I said.

"I think he *is* an MBA, and that would be all the better, Cudahy. You need some structure over there, and an MBA would maybe bring a bit of order out of the chaos you run."

"Freddie, you have such a wonderful way of saying nice things about my company that I think I'll marry you."

"I'm already spoken for, Michael, and besides I'm serious about this guy. The least you should do is meet him. Why don't you and he and I go out for dinner some time and that way you can judge for yourself. I'll buy."

"Well," I said, "I never turn down a free meal, especially from you, you cheap skate. I've met this guy before, by the way. I believe he and I were co-speakers at the Governor's Japanese/American trade conference."

"What did you think of him?" said Freddie.

"Seemed OK, I guess. I wasn't paying too much attention when he gave his talk. Kind of a little character, isn't he? I was too busy trying to figure out this guy, Morita, from Sony – he was one of the other speakers."

The dinner with Barry went fine, and this new potential president Freddie had found for me was charming enough all right, and I'd have to say he wasn't what I expected at all. He seemed absolutely fascinated with my story of how Cozzens and I started Marquette on a shoestring and how it had grown to $250,000,000 in sales in twenty-six years. He buttered me up with a bit of admiration about how I had done it, but without too much buttering. (Of course, I'd have to admit I did like hearing people marvel at what an accomplishment the whole thing was.)

Many meetings with Allen followed, and finally we had some really serious talk about his quitting Illinois Bell and what it would cost him and how we'd make it up in stock options. I thought at the time it was a bit strange this guy would want to give up the fabulous job he had just gotten. But he convinced me the communications industry was an unhappy place to be at that time in history with all the *downsizing* that was going on. Besides, he wanted to be an entrepreneur in a smaller company, and that was that. Never mind how mad Illinois Bell might be that he was only on this new job for a month or so.

And then I did what I thought would be a logical next step. I called our board members together for a sort of *grill Barry* session, asking each of the five members to really have at him to see what made him tick.

When you think about it, that might have scared the hell out of the average guy. There was Fred who found Barry and who was a super guy at talking fast and making complex business deals.

Then there was Walt Robb, a PhD in chemistry and the director of research at General Electric, and Mel Newman, our attorney from the beginning, and Gene Menden, our previous controller. He was the "historian" of the group because he could remember every mistake we ever made.

And there was old Warren, my cofounder. I've said a lot in previous chapters about this man. At the ripe old age of seventy-six, he was one of the more crusty characters on God's green earth. "What the hell would a telephone jockey know about our business, Cudahy?" he said, and he couldn't wait to meet him and rip him to shreds.

But Barry came through the ordeal with flying colors. For about six hours we asked this poor guy everything about his life from religion to sex, and I really think anyone else would have buckled under the strain.

Not Barry. He held his own like a soldier, and every one of those guys came out of that meeting saying, without question, if ever Cudahy was going to have a guy who could take the helm it would be Barry.

The long and the short of it was that Barry came to work as the new president of Marquette in the late summer of 1993.

At first, I thought everything was just great. Barry worked harder than he had ever dreamed of working at the phone company and busted his tail to do all the right things. Like spending time with almost everyone from top to bottom and laying out a strategy for everything from soup to nuts. He told me the trouble with this outfit was that we needed a hell of a lot more communication and structure, just like Freddie said, and I agreed. Never mind that the company had done pretty well for twenty-eight years, by the way.

And he decided to learn medicine and electronics from top to bottom on a crash basis. He would be tutored by every expert we had at Marquette and a few on the outside as well. I never saw a guy go at the whole thing with such explosive passion in my life. Every time I looked in his office, he was meeting with a group of managers or being tutored in medicine by Fred Robertson, MD, our medical director. And then he'd be out at the Medical College, scrubbed up and watching an open-heart operation. Made him a bit woozy at first, but he got used to it. And then he'd be up at the Research Center with his nose in some crazy, far-out experiment, asking every imaginable question.

He was something to watch in action. He was determined to take this company into the next century with more vigor than Eisenhower had stopping the Germans in WWII. Marquette was going to reach a billion dollars in sales by the year 2004, a goal I set for him in some stupid speech I made.

Maybe all MBAs aren't cut from the same piece of cloth, or maybe just most of them are. I'd have to say I know a few who are wonderfully different, but these guys are outside the bell curve. Barry was outside.

Of course what would you expect? You send a bright, young, absorbing mind off to school and pound a lot of crap into his head for six years and, surprise, surprise, out comes a guy who thinks differently from you and me. Not necessarily wrong, mind you, but differently for sure. They have been schooled in finance, and that's probably good, but then there's all this "your biggest competitor has just lowered his prices by 20 percent so what should your strategy be?" kind of teaching. Mind you, most of this stuff was generated by professorial types who have never been out there in the trenches, so half of the theories are not tested in real life. Be that as it may, MBA theory is usually a total misfit when it comes to entrepreneurial companies. Entrepreneurial companies usually have some innovation they're basing their whole operation on, so they fall totally outside the logic taught at MBA school.

Barry was an MBA, a true, dyed-in-the-wool MBA. He strategized by getting teams together and making decisions by consensus, which was exactly back-ass-wards from the way I was running the shop, and in time the troops rebelled. I gave him more rope than I ever thought I would, and I guess you'd say I gave him too much because he hung himself.

One problem with running a company like Marquette is that, in the beginning, I was a sort of benevolent dictator and that worked quite well. When a tough decision had to be made, I'd listen to every view I thought was of value and then get everyone together and say, "Well, I've heard all your arguments, and now what we're going to do is as follows."

Fortunately, I was technically astute enough and knew about what was needed out there, and reckless enough to make decisions – slightly more right than wrong. And when a tough decision was made, everyone heaved a sigh of relief that the old man was on the hook for it and they were off.

So along comes Barry, with his study groups and committees with umpteen million different views, and the decisions almost always turned to *mush*.

This, I think, is a common fault of many companies today. A committee-run outfit never makes decisions that are earthshaking so the company is never very earthshaking. It's like the old adage that when you get a bright new idea, don't tell too many people, or they'll talk you out of it no matter how good it was.

It's not surprising that Barry was very big on mergers and acquisitions. Of course merger-mania occurred, big time, in the nineties so Barry was operating at the right time in history.

He and I used to have lively debates on the advantages and disadvantages of mergers, but neither one of us ever won.

He'd say, "Take two perfectly healthy companies the size of Marquette, and no product overlap" (we were about $300 million in sales at that time) "and put them together, OK? Now, immediately, you have huge *economies of scale*. You have only one accounting department, one legal department and sometimes a whole lot less salesmen, right?"

"Yes, so?" I'd reply, knowing exactly what was coming next.

"Mike, don't you get it? You then have a better bottom line for both companies that merged. And you have a much more powerful enemy for your competitors, right?"

"Not necessarily, Barry," I'd say. "You forgot about the confusion of merging – who does what and who works for whom. The agonizing, long meetings to try to reorganize – the hurt feelings of people who think they got screwed in the deal. You forgot about bureaucracy, the number one enemy of business."

"Ah, Mike, you can *manage* all that out. That's what we executives are all paid good money for," Barry would argue.

"OK, then, why does your almighty bottom line go down with so many of these fancy mergers?" I'd ask.

"They don't, Mike. Where did you ever hear that? I can show you a million cases where they go up!" Barry would say.

Fortunately, we'd usually run out of time at that point and go back to work.

I have nothing against acquisitions if it's a logical way to broaden one's product line. But I always felt that acquisitions were sort of a form of cheating when it's only to fatten up the company. They can make the *top line* look better so people say "wow, how you've grown" but this is all done without *real work*, in my old fashioned way of looking at things. It also dilutes the original owner's equity, and with that comes less pride of ownership.

Barry, as with most professional executives, was motivated to bring about mergers and acquisitions simply because it tended to make the base company larger and thus their stock more valuable. At one time, he wanted to merge Marquette with Nelcor, a large pulse oximeter company on the West Coast, but I put my foot down on that one. Good thing I did. Nelcor was almost as big as we were so just who would acquire whom was the question. And who would be the boss was another. Nelcor was run by a very strong Irishman by the name of Ray Larkin. What an interesting battle that would have been!

Two major acquisitions did occur during Barry's reign, however. First, we bought a neat, small company called Corometrics out in Danbury, Connecticut. Started by an entrepreneur by the name of Bill LeCourciere, they made special monitors for neonatal patients. Bill apparently ran short of money some years before, and had to sell the outfit, and the company that bought him out was American Home Products, the big drug and drugstore supply company. Unfortunately, from that day on, it had been sort of lost in the bureaucracy of American. Rumor had it that the last president sent to run Corometrics never came out of his office to meet the people for two years.

Barry Allen and Mike Cudahy, 1995

After we closed the deal with American, Barry and I went down to Danbury to announce the acquisition to the troops. There were some 600 employees in the organization, and every one of them showed up for the event. We had a nice luncheon in their cafeteria and then Barry announced that we were ready to take over.

And there were hoots and hollers from every corner of the room. Liberation! Freedom! Back to the days when they were an independent company, they said! The oppression instilled upon them by American Home *dictatorship* had all but ruined this great, spirited company over the last eight years, and we were their savior.

When I stepped up to the podium I could do no wrong. I started with, "Welcome to the Marquette family."

They cheered and clapped.

"At Marquette we treat each and every employee as a member of our family."

They screamed and stomped their feet.

"And I embrace you, all of you as new and valued members of the Marquette family."

Now, things were almost out of control. The cheering and yelling and applause was totally deafening. I screamed over the crowd, "Now, I want a plant tour. Top to bottom. Which one of you out there wants to take me?"

From deep in the audience came a raised hand and a plea, "I will. I'll take you top to bottom – right now!" A heavyset lady in work clothes raised her hand and moved forward to the podium and the crowd started hooting and chanting, "Rosy, Rosy. Oh, Rosy, Rosy!"

Rosy escorts Mike on a tour of Corometrics plant in Danbury, Connecticut

Then, before I knew what was happening, I was hoisted up on the shoulders of at least six big, burly guys who took me to my newfound guide.

What a thrill! What a wonderful tour. So many were so proud of their work – work that had gone unnoticed for so very long. We had, indeed liberated some 600 people who had lived in obscurity for almost eight years.

This was not really an acquisition, however. It was more a liberation of a little company that had been mistreated by a very large company, American Home Products. And we had a similar experience with a company called E for M Corporation, which was owned by Pittsburgh Plate Glass Company. E for M had branches in California, Kansas City and Freiberg, Germany. It always fascinated me how a company like Pittsburgh, which makes glass and paint, could possibly have anything to offer the medical electronics field. And what would they know about it? It was another example of large companies on a binge to acquire – never mind what – just anything to grow the top line.

E for M (Electronics for Medicine) was the old medical electronics firm started by Marty Scheiner, a truly great innovator in the world of CATH labs, the room where catheterization of the arteries takes place. First came the X-ray imaging equipment or fluoroscopes; then, the electronics to keep track of the many human heart signals during such a procedure.

As time went on, Marty fell into bad health and the temptation of selling. So the company was acquired by Honeywell and then from Honeywell to Pittsburgh Plate Glass.

Along with the package came a grand old EKG manufacturing company called Hellige, located in Freiburg, Germany. We always felt we were not truly a part of Europe, although we had sales and service offices in many cities. So this part of the deal made us all feel like we had truly *arrived* over there. And now, we could man-ufacture many of our products within the Common Market.

Pittsburgh Plate Glass had torn down the grand old Hellige sign atop the factory in Freiburg and put up their own PPG logo when they bought it, brutally injuring the great German pride of the workers there. So the first thing we did was to have a new Hellige sign installed. They cheered. They loved us because we were considered one of them – a manufacturer of EKG machines.

Barry and I had no quarrel about these acquisitions, probably because they really weren't "acquisitions" in the true sense. As I said they were "liberations" of companies under siege. And Barry and I were in sync on many other decisions, but our fundamen-tal differences continued. He wanted a structured company run by committees, and I wanted my good old entrepreneurial dic-tatorship.

One day over lunch, Barry said, "It just isn't working, is it? – I mean, our relationship."

I agreed. "You know Barry, you've given it your best shot, but we're just cut from different kinds of cloth. I think we'd better be going our separate ways."

I like and respect Barry. Who wouldn't? I'm glad he was a part of Marquette when he was there, but the chemistry wasn't right and the team just never signed on. It was two years and Barry was out the door. He went back to Ameritech and got an even better job than when he left.

Of course, I had to go back to finding a guy who would take my place if I got hit by a truck. If this process went on long enough, I reasoned, the truck wouldn't be necessary – I was well into my seventies.

All this is living proof that the changing of the guard in an entrepreneurial company is the single most difficult task there is. Like it or not, when an entrepreneur starts a company from scratch and that company turns out to succeed, a culture is created, and the culture grows through the years until it becomes a part of the woodwork. Preserving that culture, when those who created it leave, retire or die, is very nearly impossible, yet often necessary for continued success.

Marquette was, as I said before, a product of Mike Cudahy and Warren Cozzens with the help of mentors like John Fluke of the Fluke Company, Harry Bradley of Allen-Bradley and other characters. I've described our culture in previous chapters, but here's a nine-point synopsis I wrote awhile ago:

1. Love, and listen to thy customer, above all.

2. Have fun and work your ass off but keep life beautiful.

3. Fuse personal life and company life together.

4. Make moves – man the torpedoes – even if you're wrong.

5. Don't worry about the competition – just do the job.

6. Innovate – try anything that's new and might work.

7. Don't get too structured – keep the pyramid low.

8. Treat everyone in the outfit with respect and love.

9. Be serious but humorous at the same time.

Barry was gone and the problem had to be addressed. I was dead set against another outsider for a million reasons. The process of *headhunting* and the so-called professional headhunters make me gag. They look around and try to find people who are either unemployed or disgruntled with their present job, and that's a bad place to start. If some big shot is unemployed, why? If disgruntled with his present job, he's telling people on the outside about his frustrations? Behind his boss's back he's negotiating with some headhunter to get another job? Do I really want that kind of animal?

Even if the headhunter finds somebody who looks really good, how can you tell? Past record? People used to call me all the time and say, "Bob Smith used to work for you, didn't he? I'm a headhunter calling to get your candid opinion of this guy, and, don't worry, Mr. Cudahy, I'll hold anything you say in strict confidence."

Sure you will.

Even if I believe the guy and spill my guts out about Bob and that he was a drunk, a thief and stupid, what am I supposed to do, ruin his chances to ever get another job?

And even if I find what appears to be a really good candidate, and, yes, first impressions are supposed to be the most accurate, and I follow my gut and all his credentials are fantastic, my chances of really knowing this stranger at time zero are next to zero. It's going to take at least six months to find out what the guy will do in various circumstances. And then, if he's not working out, I really should tell him why and give him another six months to see if he can change his ways. And if he can't or doesn't want to, I have to start all over, having blown a whole year.

No, I said, our new president must come from within and so, insider Tim Mickelson was the next logical choice. He had a PhD in physiology, was liked and respected by everyone in the company, and had been in and out of Marquette for many years – meaning he joined us in about 1982 and then quit for awhile to join another company, but came back a few years later.

When he rejoined Marquette, he ran the Monitoring Division for about four years. After that he took over our

Corometrics Division for two years, the company we bought from American Home Products that was in neonatal monitoring.

Tim was everyone's next choice for president. One thousand people at our Milwaukee plant signed a congratulatory plaque when he got the job, which would lead you to believe Tim was one very popular guy.

Popular or not, Tim was too placid. He never seemed to get exercised over anything. He never swore or screamed at any-

Tim Mickelson

one, and it looked to me like he would go to any lengths to avoid a confrontation. (I never did much screaming, either, but once in awhile you have to let people know you're capable of it.) Tim was what I called a "Tan Buick Sedan." Nothing wrong with tan Buick sedans – they provide good transportation for millions of people every day – but Marquette needed a sports car.

Tim made what seemed to be logical decisions, or sometimes put off decisions, and things started to get very boring around old Marquette, yet nothing was really wrong, either. He apparently decided the way to handle the old man was to avoid him. Just keep everything smooth, and maybe Cudahy won't ask too many questions. And when I did ask, I was always told "not to worry, everything's under control" no matter the subject.

One day I was bitching about the performance of one of our divisional presidents, and Tim was trying to tell me the guy was just fine and everything was OK.

He said, "You know when you get it in for someone, it's just a matter of time before he's out the door." He went on to say it was just a matter of time before I would get rid of him as well. Sort of mumbled it, but I'm afraid I heard it clearly, nonetheless.

"What was that, Tim?" I asked. "Do you really think that? And if you do, why in hell do you stick around here in this job?"

"Well, I have a wife and three kids for one thing," he said in a totally calm and quiet voice.

There goes my "everyone loving their job" theory. I really felt sorry for Tim, but what I needed was a big, red Ferrari.

CHAPTER 12

THE DECISION TO SELL

I n 1996 our stock was, so they say, too *thinly traded* to be very healthy on Wall Street. What this means, at least in my understanding, is that there just aren't enough shares to go around, so big investors (like institutional investors) don't like to get involved. It results in a stock that, when you go to buy very much of it, you drive the price way up and then, when trying to sell it, you drive the price way down.

We were doing very well at the time and gaining each day against all of our competitors. Companies like Hewlett-Packard and Spacelabs Medical just couldn't fathom how we did it – perhaps they should have known it was just plain hard work.

Yet our stock was lackluster. It was trading around 16 – 18 at the time – about twelve times projected earnings. High tech or not, we were in the gutter.

One way to solve the problem, I was persuaded, was to have what is called a *secondary offering* – to sell more of my stock, the company's stock or any inside holdings so as to increase the amount available for trade.

And what happened next was to change the course of our company forever. It would cause me to rethink my rigid stand about ever selling the company. These next events, traumatic as they were, made me realize that nothing was forever.

In preparation for this so-called *secondary offering*, we visited with a few investment bankers in New York to get some advice from the pros. When we had our first public offering of Marquette stock back in 1991, we had the job done by a local firm, Robert W. Baird, a division of Northwestern Mutual, the "quiet company" as they advertised. They did OK in their quiet way at first, but over the years they grew tired of our stock and found other fish to fry.

We contacted one firm because its president, Fritz Hobbs, was a friend of our president, Tim Mickelson. The company was Dillon, Read, and Fritz was the fastest talking character I have ever met. We Wisconsin farmers used to call this kind of guy a "Big Butter'n Egg Man."

Plans got underway in December to file our registration with the SEC, and an official roadshow was planned to start in February, 1997. Mickelson, Mary Kabacinski, our chief financial officer and I would go with some of the guys from Dillon, Read, and the show would start in Europe. Then we'd do the West Coast, the Midwest, the East Coast and wind it up in Texas and Canada. Three weeks of grueling travel, living on a plane, climbing in and out of long, black limousines (for some reason, Wall Street loves limousines – the bigger and blacker the better) and checking in and out of pretentious hotels. Three weeks of telling the same story over and over about what a fabulous company we had and how we were looking to the future with expectations of greater sales and greater profits.

During the last week of this lovely experience, we left Milwaukee on a Sunday night for Dallas. The next morning we had a breakfast meeting and a midmorning meeting and then went off to Austin for a luncheon. After lunch, we took off for Houston for, you guessed it, another meeting, and as soon as we were finished, we ran for the airport to make a dinner meeting in San Antonio. After dinner we had to rush because we needed to get to Toronto in hopes of catching a few hours of sleep before our 7:30 breakfast meeting up there. At seventy-two, I was feeling pretty smug that I could hang in there with Tim and Mary, who were in their forties.

In pricing the stock, I was the guy officially designated by the board to give the OK because I had a million shares in the deal of my own. Besides, I was the chairman of the board and close to all other members and their opinion of what was fair and what was not. To sell new stock in a *secondary*, you don't necessarily have to stick to the price of the stock that's already out there, I found out.

Freddie Luber, our most outspoken old-time board member, thought that we should not price the stock for a penny less than $19, and he wasn't even happy with that. His parting words to me before the big day were, "OK, I'll go along with whatever you decide, but I think if it's any lower than $19, we should just wait 'til sometime later. That stock, Cudahy, should be something like $25 a share so what's the rush to sell cheap?"

The other board members felt pretty much the same way. I remember Mel Newman, also a long-time board member and the company attorney, saying, "I don't know why everyone is in such a hurry. I'd wait. If the price isn't fair, just cool it for awhile."

I wasn't about to tell our little road show gang where the Board's *go-no-go* price point was. But in my mind it was $19, and the thought of my telling the Dillon, Read company their suggested price, whatever it was, might not be good enough, sent a chill through my poor old bones. If that happened, the entire road show would be wasted. The market price of our stock being traded at the time was slipping as we approached *pricing day*, and all I felt I could do was hint that we might not have a deal.

An owly-eyed young guy from Dillon, Read by the name of Dave was along on most of the roadshow. In the last few days, I said, "Now Dave, we are the ones to make the final decision on the price, are we not? That is, if we don't like the price you guys suggest, we can say 'no' can't we?"

"Of course, Mike," Dave would say. "After all, you are the ones who are doing the selling, and you never have to sell anything you don't want to sell, do you?"

"That's good, Dave, because the trading price out there is barely $18, and we may not have a deal if things don't improve, you know."

On the big *pricing day*, I was back at home and others were in different cities, so the process was staged by conference telephone. I'll bet there were fifteen or twenty lawyers and brokers on the line in New York, Chicago and Milwaukee. The brokers had given me a rundown on the day before – a summary of who had made *pre-pricing* offers for the stock so far.

Some of the offers were as low as $17, more than a full point below the market price, and I told them that was ridiculous. On pricing day, Mr. Butter'n Eggs was saying that he had "talked those low ball guys up to $18 and that was a miracle." I said that was nowhere near good enough.

It didn't take long for someone from Dillon, Read to announce to the group on the phone that the market price was $18 3/8, and that was how they wanted to price the new stock. I said that I regretted that because my answer had to be "no sale."

The shit hit the fan! First to speak up was Butter'n Eggs with a tirade I wish I had recorded. He was almost out of control and told me this was the damnedest thing he had ever heard – my not accepting such a good deal. He said 18 3/8 was the market price and that was the only price they would agree to, and I had to go along with that price and that was that!

I waited until the ruckus quieted down a bit and then said, more or less calmly, "I told your man Dave a hundred times while we were on the road, that we may not have a deal at such low prices. He assured me it would be my option to say 'no' if I so desired and at this suggested price, I'm sorry, but I have to say 'no'."

But the screaming started up again, as many angry voices chimed in from around the country. The price of $18.37 was fair. I had to move forward. I just couldn't say "no." Then it was agreed that Mary, Tim and I would have a separate conference and reconvene in an hour. I was relieved and hoping that in that hour they'd cool off and come up with some "sweetener" closer to $19.

Mary Kabacinski, our chief financial officer

Soon I was hooked up with just Mary, Tim and lawyer Mel. Mary was upset, disappointed, but very professional about it. She agreed that it was my option to say "no" and that I appropriately spoke for the board of directors.

Then Tim chimed in with what started out to be a reasonable speech to try to change my position, but soon he got carried away and, as they say, lost his cool. He was some distance from the phone so it was hard to hear, but the essence was, "This is ridiculous – I quit!" as he was going for the door and slamming it with all his might, which was considerable. Mary said, "Oh, this is terrible. Why would he quit?"

Then Tim, unfortunately, found a pay phone somewhere on the street and called his friend Hobbs, announcing his resignation.

If I ever thought I was going to drive the price up to our desired 19 by saying "no," there went all my bargaining chips down the tubes. Even a quarter point meant almost three quarters of a million dollars for those who were selling, including me, and I really hoped Hobbs would nudge up the price so I could satisfy my board.

I called the brokerage group back and told them we would reconvene at 6:30 a.m. the next morning after everyone had had a chance to reflect on the situation. But I knew I had little ammunition, if any, left to play any hardball.

I hung up the phone and bent over in my chair, dropping my papers to the floor. I felt tears on my cheek.

Where were the good old days? What happened to my wonderful company? Where has the joy gone? This isn't Marquette anymore. This is the ugliness of Wall Street and all its crass commercialism.

I felt old for the first time in my life. Very old. The *Joyworks* I had created was now clearly behind me. The future of Marquette would never be the same. Now it would only be about money. About trading stocks. About mergers and acquisitions. Forget the fun of conquering new technologies. Forget the wonderful employees we have had through all these years. Forget the great doctors and nurses who had become our best friends.

"It's time to sell," I whispered to myself. "It's time to hang it all up. It's time to find some other challenge for the rest of my life."

Then I remembered the Harry Bradley thing — when he would come to our house and tell my father his intention was never to sell his company, Allen-Bradley. But Harry died and Rockwell International came in for the *kill* and scooped up the company for a cool $1.65 billion. Harry could do little about it, being six feet under the sod. It convinced me that nothing was forever in this world.

So sell the company, damn it!

The next morning we reconvened en masse on the phone, and I announced that I'd go along with the price they had offered. Old Butter'n Eggs gladly agreed and the deal was closed. Tim agreed to come back and work hard to make things OK again although I can't imagine how he rationalized the whole thing. I figured that if I were going to sell the company, it would be a hell of a lot easier if I had a running, working outfit – including a qualified president at the helm.

The board agreed. It looked like the time had come to at least look for a company that might be interested in acquiring Marquette. My old, trusted employees, I reasoned, would probably rejoice if I sold out. After all, they were all getting on in years, and the cash from their ESOT (employee stock option trust) would probably be a welcome surprise for their retirement.

Tim Mickelson only lasted a year after all that ruckus, and when he resigned, I was really at a loss as to who could step in, when I suddenly remembered our former medical director and one-time head of our Monitoring Division, Dr. Fred Robertson. Fred had left the company, partly because of differences with Tim, but also because he wanted to get back into medical practice. He was doing well as head of the surgery unit of the Milwaukee Medical Clinic.

Could I possibly talk Fred into leaving the clinic and coming back to Marquette? Turned out that Fred, too, was wanting to come back. And for that last year before we sold to GE, Fred turned out to be the best, not *clone*, but a guy who could really carry the ball in our crazy company.

CHAPTER 13

GENERAL ELECTRIC

I had been working on a project with Johns Hopkins Medical School for some time to develop a system cardiologists could use in getting images and data on their cardiac patients. Our aim was to get EKGs, stress tests, Holter, MRI, CAT scans, and chest X-rays all on the same computer screen in the same room – and all more or less simultaneously.

I realized we would require a lot of help from other companies. Marquette knew EKG, stress and Holter, but this system was to extend way beyond these tests into fields we knew very little about. We needed experts in X-ray, CAT scanning, and MRI.

General Electric naturally came to mind. GE was number one or number two in all of these technologies, and the Medical Division was in our back yard (Waukesha, Wisconsin). It was only natural that I talk to GE about the project at Johns Hopkins. Jeff Immelt was the new boss of that operation so I gave him a ring.

"I was going to call you, Mike," he said. "I know about your Hopkins project, and I think it's very exciting. Let's get together over lunch. There are a number of other things I'd like to talk to you about, too. Name your place."

"Know where the River Lane is, Jeff?"

"No, but I'll find it. How about noon on the fifteenth?"

And on that cold 15th of January 1998, Jeff Immelt and I gazed through the restaurant window at the falling snow and started a casual conversation about Johns Hopkins. But one of the other things Jeff wanted to talk about was the possibility of putting our two operations together in some sort of merger. I suspect he used the expression "merger" because he probably thought the words "acquire" or "buy" might rub me the wrong way.

He was right. After thirty-four years of growing Marquette, the hard, cold thought of selling my baby sometimes made me squeamish, even though the board and I mutually decided that was the best thing to do. Somehow, Jeff knew how sensitive I was about it, so he tactfully said he thought a merger might be good for both of our operations. And after touching the subject, we got some coffee and went on to talk about the Johns Hopkins project.

"Let's have lunch together again, Mike. We really have a lot to talk about. I'll give you a call," he said as we bundled up and braced ourselves for the cold.

But Jeff never called. The winter went by and I was determined not to show any eagerness by calling him even though my board thought maybe I should. Finally, in July, a half year later, I rang Jeff up and asked if he was still interested in the things we had talked about in January. I added that other companies were starting to talk to me about acquiring Marquette.

That must have got his attention. He said "maybe we should have lunch again – soon."

In the six months that I heard nothing from Jeff, I happened to meet an interesting guy by the name of Dr. George Hatsopoulos, the founder of Thermo Electron Corp., a large conglomerate of medical and other electronic companies. He and his biomedical CEO, Jack Keiser, expressed a strong interest in folding Marquette into their group so our president Fred Robertson and I went to visit their world headquarters in the Boston area. Unfortunately, the stock market then clobbered Thermo because of a problem they had with one of their subsidiaries. Thus, acquiring Marquette was probably out of their reach.

Then there was Hillenbrand Industries, the people who make most of the hospital beds for North America. Gus Hillenbrand, a fellow airplane owner and friend, let it be known through one of our investment bankers that he'd like to talk to me. Trouble I had with that was that the other half of their business was making caskets, and that sort of bothered me. GE seemed a much more cozy and sensible idea if we were really going to go through with this thing.

The phone conversation with Jeff sounded as if GE still was interested so the next lunch was likely to be pivotal. I had to be careful not to let Jeff know how interested we really were because the next topic might be how much they'd pay for Marquette. And I'm sure he had to be very careful to make sure I didn't think *they* were that interested. Silly, the games businessmen play.

Then I heard through the grapevine that GE was thinking about offering "in the high thirties" per share for Marquette, and I was almost inclined to forget the whole thing. Surely, we were worth more than that! Other companies were doing much better in the sellout game, like Physio-Control, a one-product defibrillator company we knew well and thought little of. They were bought by Medtronic, the pacemaker company, for a totally exorbitant sum. So what the hell was wrong with us?

Jeff finally got around to talking price at that next lunch but seemed to think GE would *never* pay a penny higher than $40 per share. "Better than the high thirties," I said, "but I wouldn't even consider mentioning it to my board for anything below $50."

On the way out the door, Jeff said, very softly, "Maybe the solution would be to see if we could split the difference," and I answered, "that would still be $5 below where I would consider *any* offer." I added that maybe it was time to take a run down to

New York and talk to Jack Welch, the chairman of GE and, of course, Jeff's boss.

"Maybe you're right. See if I can set something up, Mike," he said. "I'll let you know."

On Friday, September 11, 1998, Jeff and I took off in my Hawker jet for the big city to see Mr. GE, Jack Welch, in his office atop Rockefeller Center.

In the short time I spent with Jack, I loved the guy. Humorous, sharp and very fun to talk to on any subject. He came roaring out of his office as Jeff and I approached and said, "Hello, Jeff. Say, you must be Mike Cudahy."

Usually meeting fellows like that involves several layers of secretaries and checkpoints and glass doors. Usually, the guy sits in some huge swivel chair he spins around to intimidate you when you're ushered in. Jack was right there in the hallway with a friendly handshake and said, "Come on in. I've got a little light lunch for just the three of us."

The room was small and tastefully decorated with a round table just big enough for the three of us – no papers or contracts – just a simple lunch.

I said, "Welch. Welch sounds Irish to me."

"100 percent Irish, Mike, and I'm proud of it."

"Impossible," I said to his surprise. "No one is 100 percent Irish."

"What do ya mean?" said Jack, with a little hint of indignation. "My family came directly over from the old country, no stops along the way. Why, I'm Irish as Paddy's pig!"

"Can't be, Jack," I insisted, "The Irish are all mutts. There were the Normans, and the Vikings, and the Spaniards and don't forget the English. Now, you don't think in all their invasions of our fair land they didn't have a bit of fun with our ladies, now do ye?"

Jack was taken aback. But he looked at me with a pair of powerful blue eyes and said, "I guess you have a point, Mike, but mutts are the best kind of dogs anyway, don't you think?"

"You bet they are. And *my* family is no exception. I even have a wee touch of Scotch in me, so what do ye think of that?"

"No one's perfect," he answered with a warm smile.

"I'd like to know if the lunch looks OK to you. It's just a simple salad. What would you like to drink? Coffee? Let me get you some."

With that, he grabbed what resembled an old-fashioned garage door opener from the table and pressed one of the buttons. Soon, a slightly built Asian-looking man appeared, and Jack told him to bring Mr. Cudahy and Mr. Immelt some coffee.

"Yes, sir," said the man and went back behind a screen that separated part of the room.

"How many kids have you got, Mike?" Jack asked.

"Five, last time I counted."

"Any in the business?"

"Well, yes, my daughter Julie is an engineer and is buried somewhere in the outfit. The rest are scattered around the globe doing things unrelated to Marquette. I have a son, Patrick, who's a commercial photographer out in California. Doesn't make any money at it but he's happy as a clam. You know, Jack, I finally figured out in my old age that everyone in this world, especially our kids, doesn't have to be a fanatic achiever – like you and me."

"Funny you should mention that, Mike. I have a daughter up in Martha's Vineyard who drives a bread truck. She's just like your son, Pat. She's happy as a jaybird up there doing her thing. The hell with old dad who's chairman of GE. Rest of your kids?"

"Susie's married to a Fed-Ex pilot, and Mary to a farmer. Mike Jr. is starting a business venture of his own," I said.

"Mike, I see this merger with your outfit and ours as a true win-win," Jack blurted in an abrupt change of subject. "We have a great medical division, and you run a fabulous medical electronics company. If we put 'em together we'll have a world leader that'll be hard to beat. Only problem is that you want too much for your outfit, and we can't spend that much for it. You talk about $50 a share, and we're talking $40. Forty comes to a hell of a handsome multiplier, you know. Your stock is only selling at $26 and we're kind of stuck on this point. Next thing you'll ask me is if I want to flip a coin. You say $46 and I say $44."

"I've got just the coin, Jack," and I reached in a special compartment of my wallet and produced a 1926 Irish half-crown. Jack looked at it and said, "Hey, let me see that. Nice coin. But wait a minute. I can't flip. I had a board meeting this morning, and they authorized me to offer you somewhere between $40 and $45, and that's all I can do without reconvening the board, and you know how impossible that would be."

"Well, I guess that means that you're offering me $45 a share."

Jack dropped his napkin and was looking for it on the floor.

"I guess you're right," he said, coming up from below. "God, the board is gonna kill me for settling all the way at the top of the range."

We reached across the table and shook hands. A deal for almost a billion dollars had been struck in that instant. And there was never a question it would go through because Jack and Mike shook on it. The details would be worked out by others.

"It's worth every penny of that, Jack. You'll never be sorry," I said, and all Jeff could add was "I guess we've got a deal."

That simple handshake touched off a flurry of activity by executives, accountants, lawyers, investment bankers, and staffs of the two companies all over New York, Milwaukee, and Waukesha.

The mechanics of the job must now be done as quickly as possible, yet every move of every person involved had to be kept totally secret from the outside world. The old military adage *need to know* had to be applied in its strictest terms.

I conducted all of the meetings between the companies at my residence in nearby Cedarburg, Wisconsin, so this contingent of s*uits*, as my ex-wife called them, accountants, lawyers, and the like, wouldn't be seen around Marquette's main headquarters. We all knew that if the agreement were to leak to anyone outside the circle of workers directly involved, it would present a phenomenal opportunity for someone to make an illegal fortune in what is generally called insider trading. Anyone with the knowledge that Marquette stock, then trading at around $26, would soon be transformed to GE stock worth $45 per share was sitting on keg of dynamite.

All that week and on into the weekend, the *suits* drove in and out of our long, winding driveway and used every telephone in the house, checking their contracts, verifying their numbers,

Jeff Immelt and Mike Cudahy sign the sale of Marquette Electronics to GE.

making sure their counterparts on the outside agreed. When the lines were busy, they each had their cell phones. You could find them in the bathrooms, bedrooms, hallways and out on the lawn, headsets crammed to their ears, antennae extended.

On Monday, September 21, just ten days from that famous handshake, I delivered a speech to our troops in Milwaukee with a TV hookup to our plants in Florida, California, Wallingford, Connecticut and Freiburg, Germany, announcing the deal with GE. I wasn't too sure how they'd take to all this but now was the time to tell them, for better or worse. I reasoned that a lot of the gang were getting on in age and would realize that their ESOT (employee stock option trust) would now be worth almost twice as much as it was the Friday before.

The speech went:

> I wrote the *Sellout Syndrome* in 1978 (*see page 85*). It lamented how so many small medical electronic firms were selling to the big drug companies. It said that their efforts and successes should be more important than just money.
>
> Today, it's a slightly different picture. We're not "small" anymore. We're almost $600 million in sales, we're a public company and a lot of the original Marquetters have retired or are getting ready to retire.
>
> And I'm sure you'll remember I also wrote an article for *Fortune* magazine entitled *Going Wrong by Going Public* back in 1984 in which I said a company should be all grown up before going public.
>
> Well, we went public in 1991 because, I guess, we decided we were grown up, and we had to give our ESOT shareholders a tangible value for their stock before they retired.

So when we did go public we issued a special *C* stock in order to protect all of us from what's called hostile takeovers. With that I had 84 percent of the total vote and could tell anyone where to go – and, of course, I did.

But in 1995 I turned in my *C* stock because Wall Street told me it was depressing our stock price and, sure enough, the move did actually cause a slight gain in the market.

Now, I can tell you today that we have had several suitors who have expressed a strong desire to acquire the company, even a hint by some that they could take over if they wanted. If they did, you realize, they might easily ruin our culture and strip the operation to provide themselves with more profits. Some might, for example, close our day-care center.

And as a result, the board of directors said that we should entertain the idea of *merging* with someone, not only to end the takeover threat, but someone that might fit well with us and our products and that could also preserve our culture – if that were at all possible. Someone that could help make us a truly world power in medical electronics.

Well, one such company answering to all of these requirements is GE Medical Systems, located right here in our backyard, Waukesha.

Remember GE? That's the outfit that wanted to buy Marquette in 1982, but, instead, we bought a part of them. I mean we bought the monitoring product line of GE and turned it into a huge success.

So when I had a talk with GE Medical's boss, Jeff Immelt, some months ago about all sorts of cooperative efforts with our mutual customers, he surprised me by saying they had a strong interest in merging with Marquette.

I told him if such a thing would ever happen, we would insist on preserving our culture, our style, etc. and Jeff not only agreed to that, but assured me that this would be in their interest as well as ours. And we both believe the synergism between GE and Marquette is without question, particularly in cardiology, our fundamental business.

Yesterday we signed an agreement to merge with GE Medical. All this has to be approved by the stockholders and the *feds* of course. But if all goes well, we've got a deal with GE Medical Systems which I feel will be good for our company, good for our customers, our stockholders, and most importantly, good for our employees.

And I believe today that, in spite of the famous sellout syndrome I wrote those two decades ago, that an affiliation of this sort is absolutely necessary to cope with today's market, and is necessary, as well, to assure our continued growth.

I like the people at GE. I like Jack Welch, the Chairman of the entire operation, and I especially like Jeff Immelt, the President of GE Medical Systems. And yes, in spite of the fact that money isn't everything, their offer of $45 per share might just be received by our old-time ESOT-ers with a certain amount of joy.

Chairman Jack Welch has asked me to stay on for at least four years, an offer I will be quite flattered to accept. That means you'll still have me around (for better or worse) as well as Fred and many more of the old *regulars.*

I meant every word of that speech. And the original intent of Jack Welch and Jeff Immelt *was* to preserve our culture, totally, in their words. Immelt said at the time, "It's a going, profit-making, successful company. Why should we mess it up?"

I wrote in my next column of "The Beat," our company newspaper, as follows:

It's hard to think of where to begin writing for this
issue of The Beat with all the excitement about the GE
announcement last Monday.

First, I'm happy to say I didn't see too many long faces
as I walked through the Milwaukee plant and visited with
many of you. The old timers were particularly happy, and
why shouldn't they be with their ESOT value suddenly
almost doubling. I'll be visiting other sites just as soon as
I can.

There is, understandably, some anxiety about *will these
guys really not change the place* and, of course, job security
with a few, but as I have told everyone, we wouldn't have
done this thing if we didn't think it would improve business, and improved business means more jobs, not less!

But a lot of people wanted to get into my soul, so to
speak, about how I really feel and I'm flattered that they
care. I'll try to be as candid as possible with you on the
subject because, as you know, I usually say it like it is.

You can imagine after thirty-four years of doing this
thing, it's a bit emotional, but it really had to happen, and
now is the right time in my opinion. And I don't have a
drop of seller's remorse about it. I think we did the right
thing for everyone in the long run, and I like the gang at
GE – a lot. I shudder at the thought that someone else
could have bought us against our will and then what
might have happened to our good old Marquette culture.

I told Jack Welch, GE's overall chairman, that I wouldn't
have had much enthusiasm about the idea of GE Medical

and Marquette a few years back, but that President Jeff Immelt is a guy I can relate to and has brought about some great changes since he took over. Jeff thinks a lot like I do, and Jack Welch pretty much lets each division president do his own thing. Sure, nothing is forever, and I suppose Jeff could leave and someone new could come in, but I think he's in it for the long run and loves what he's doing. He and Fred are getting along just fine, by the way.

For those who missed last Sunday's *Journal*, they were pretty much on the mark I thought when they said:

> **MARQUETTE CULTURE STAYS, SAYS GE …**
> General Electric Co. had a message for its newest employees last Monday morning: "We won't spoil the fun at Marquette Medical Systems." Preserving Marquette's corporate culture – including casual clothes and beer for lunch – was a key point in the company's $900 million deal, said Charles Young, GE Medical Systems spokesman. "If people want to drink beer and wear shorts to work, more power to them, as long as they continue to make the best cardiac and patient monitoring products in the world," he said.

But saying these things and doing them were two different matters. To start, Welch was very, very busy running his $100 billion company, GE, one of the most successful companies in the world.

The moment I left Jack's office he was off in nine thousand other directions and so was Jeff Immelt. Jack said, "I'd like to come out and see your operation someday, Mike," and I said I'd like to give a little dinner party for him with some of the top executives of Wisconsin's leading companies, but it never happened. Immelt said, "Let's plan to have lunch or dinner at least once a month," but that only happened twice after the sale.

Jeff Immelt is a big, humble, likable, capable guy. Originally pursuing a professional career in football, he had that easy, fair play style often found in athletes. He is the kind of guy you'd expect would have had experience in selling, and he did. Before the Medical Division, he ran sales and later management for GE Plastics Division.

His schedule was out of some bizarre movie script about the ultra busy executive. He wanted to help with the transition, but was running a $5 billion division of his own, GE Medical. He had no choice but to put a half dozen young MBA types in place at Marquette to help our then CEO, Fred Robertson. Trouble was, these guys were totally inexperienced in sales, marketing, people skills and the culture they inherited. They proceeded like the original bulls in the china shop, and the old gang – the people who really made the place what it was – didn't like it one bit. The innovation part of our culture was scrapped in favor of better bottom line performance. The warm and cozy atmosphere we were so famous for blew away in a matter of months.

I wonder how many of us really knew what we had at Marquette. Customers had a vague idea. The magic of a team of thousands, all playing the same tune in the same key. We were such a finely tuned orchestra that we could nudge each other and know what the other guy was up to. Many of us had been together for ten, twenty and even thirty years refining this intercommunication, love and understanding, yet everyone was free to "solo" (innovate) if he or she wanted to. We were like the Count Basie Jazz Orchestra in, say, 1978. We had a silky smoothness in our operation that our competition simply could not fathom.

When GE took over, they couldn't even begin to understand such a culture. They effectively said, "No more of that jazz. We're going to play pop music because it will improve our bottom line, and you folks are going to love it."

And I could hear the gang saying, almost in unison, "Oh, shit" – and wondering where they would go for a new job.

Four of these young GE MBAs asked me to have dinner with them one night so they could get a handle on this strange culture that they inherited. I said "sure" and was flattered and encouraged with the request. Were they really going to listen to the old man and try to understand Marquette?

The dinner was a bust. Mike, Brian, Charlie and Steve listened politely at first, but couldn't really understand. When I started to explain why we didn't have time clocks they nodded politely, yet you could see by their faces that they were saying, "Who's he kidding?" When I said trust in employees was more important than their pay, I started losing my audience.

How could I have expected anything different? These were bright young graduates of business administration schools who had been taught totally synthetic examples of case histories in business, invented by university professors who never experienced the real world – who never had sold, negotiated or quieted the wrath of an angry customer.

And I wondered why? Why would GE pay almost a billion dollars for a company they admired and wanted to own so dearly and then blow it all away? The good news was that in spite of the destruction of the company culture, our 200 person sales force was suddenly augmented by 800 eager GE salesmen all over the world. Thus, sales were up dramatically in the first years.

As time went on, it became more and more difficult to know what really was going on at Marquette. We were buried now in a $100 billion company. GE bragged that they were making money, but I doubt that there was a true accounting of the immense additional overhead. One thing I did know was that our trusted and loyal employees left in droves, unable to put up with the new management style. In fairness, however, the old-timers were, perhaps, nudged along by cashing in on their GE stock. We created quite a few millionaires, fifty-seven worth over $500,000 and hundreds with over $100,000. But many, many of the younger generation without much financial incentive departed as well.

I talked to lots of the old gang over lunch or in the hallways. The story was always the same. "Gee, Mike, things just aren't the same." And I got many touching letters. Some samples follow:

From Teresa Guichard

I have a few things for which I need to thank you before I go to look for another job.

You probably do not know who I am, so here is my brief introduction. I was one of the Engineering Department mechanical designers in the Supplies Division. I am extremely shy and have problems expressing my feelings. That is why I did not have the courage to say thank you before. I was working for Marquette for almost nine years. This place was one of the greatest companies I could work for. I always felt like being a part of it. I hope that my next place will be at least half as good as Marquette. Thanks for the memories.

From Wyndham Gary, Jr.

Thanks for some of the best working years of my life. A day doesn't go by since the acquisition that I don't think of how great the last twelve years have been. Your simple ten rules ring in my ears, along with the pride you created in every product that was Marquette. I know now how hard it is to create a winning product, culture and team, and how easy it is to lose.

Thanks for your belief in engineering.

Thanks for teaching us to listen to our customers.

Thanks for your understanding with respect to product schedule delays.

Thanks for your ability to see everyone working for you as an individual.

Thanks for letting us work as a team to beat our competition.

Thanks for your heart.

From Hilary Cellard

I just wanted to drop you a short note to say goodbye and to thank you for creating a company that I have really enjoyed working at for the last eight years. It is time to move on, but there are many memories from my stay here, which I will cherish forever. You are a great, charismatic leader, and that was always a motivating force for many of us, even for those of us here in Europe. The fact that you remembered my name and the fact that I worked in the Paris office when I dropped by the plant in Milwaukee was something which really meant a lot to me.

All my best to Lisa and to you and JoAnna in your new and exciting endeavors.

From Wesley Davis

Well, like the song goes ... "all good things must come to an end." I have procrastinated writing this final e-mail to everyone, but I promise I'll try to keep it short.

While I am eager to take on the new challenges over at Health+Cast, there is great regret in leaving the Marquette family. I feel very fortunate to have been a part of a great organization and to have shared in the success. There is no doubt that Marquette's richest asset and greatest contribution to the health care industry was the people – the individuals at Marquette who worked tirelessly on project after project, and often made seemingly impossible tasks both enjoyable and rewarding.

Over the years, I watched the company attract and retain a wealth of talent. Most everyone came to Marquette, not necessarily with dreams of glorious wealth, but with a certain feeling that employment by this organization offered the opportunity to grow professionally, without the boundaries and bureaucracy typical of other companies.

I will not forget the friendships and the many great times that I enjoyed. Goodbye, Marquette, and best wishes to everyone.

From George Kriegl

I enjoyed being on the technological edge and being allowed to "push the envelope" in applying new technology to our products. I enjoyed the casualness of the dress and the lack of rigid hierarchy/bureaucratic structure as well as the flexibility of hours and schedules with the lack of time records or the micromanagement of my activities by some overseer.

My children, David and Danny, have warm memories of their time at Marquette and I know that they hope to have as rich a work environment as their Dad had when they enter the work force.

The plays put on by Marquette were a real treat. I enjoyed them in a special way because the actors were our friends and coworkers. The engineering parties and nights out were a truly enjoyable and effective way to make the design teams cohesive and bonded. I enjoyed the landscaped grounds that Charlie kept up so well, the ability to borrow the company van on occasion and the fact that I could buy a number of items through Marquette at a discount for my use.

From Charles R. Smith

These are just a few of MJC's "golden rules."

Don't play big shot. No mahogany row. No special parking places or entrances.

Let people do their thing.

Give away at least some of the store.

Don't make an organizational chart.

Don't have meetings.

Don't hire professional managers or consultants.

This philosophy worked. MEI took the broken down GE monitoring products division and turned it into a $500 million dollar market leader. It was done without Six Sigma, Work-Out, career bands, Session C (code 4's for everyone) and micromanagement ad infinitum.

It was trying and failing and fixing and doing it because you wanted to succeed as a team. We had one objective, *Customer Satisfaction*. That meant listening to them and involving everyone at MEI in solutions. I learned "Just Do It" from MEI not from a Nike commercial. We are talking about innovation. The list of firsts and products that worked is worthy of a medical device hall of fame.

I can say along with hundreds of other Marquetters I was there. I worked with the finest group of people at the best company in the USA.

I am forever grateful to Mike Cudahy and Warren Cozzens for their vision, strength and hope they gave us all.

From Priscilla Tentoni

I look back at what a wonderful place this was and a part of me feels so very sad.

Just last Saturday, I was talking about Marquette with a friend. I told her that I have decided to feel grateful for having had the opportunity to work at a place that I really loved, even if that place is no longer a reality – just a very fond memory.

Some people work their whole lives never having the opportunity to work at a place like Marquette. I've decided to just feel grateful for the memory.

Shortly after the merger, I wrote a letter to Jeff Immelt, GE's Medical Division boss, about what was happening. It read:

A word from a worried old man.

It's now been awhile since the GE Marquette closing, and I'd like to give you some observations as to how things are going – from my point of view.

First, the GE people I've encountered have been courteous, smart and competent. Not bad for a starter. And they have every intention to make things go well.

But what's happening to the old Marquette drive to conquer the world is something else. There have always been a handful of key people in our outfit responsible for much of our success in winning big orders and, frankly, these people have been neutralized almost completely by an endless string of meetings. I mean guys like Fred Robertson, Gary Close, Lou Scafuri and many more are so busy attending meetings that they have literally been deactivated – taken out of the field completely.

And customers are suddenly not coming to the plant, the demo room, to look and see and buy and talk to these key people. I don't know where they are, but I suspect they are not coming because they're not getting that old Marquette "come on, you've got to see our operation" treatment that always got them there in the past.

Some day everything and everyone will be organized – when the meetings are over – and they'll look up and say, "where did all the customers go?"

I'm really worried, Jeff.

Jeff never responded.

Looking back at GE's culture vs. our culture it's now obvious to me how totally different we were, how completely incompatible. Not that one was good and the other bad – just different. Jeff Immelt wrote five "objectives" for the future of Marquette shortly after the acquisition. They were:

1. Accelerate Asia Growth
2. Global Cost Competitiveness and Facilities
3. Service Supplies Integration
4. Information Systems Strategy
5. Top Line Cross-Line Pull-through

What was he saying? And where did he say "continue to innovate" in these objectives? Where did he say "preserve the culture" we so laboriously created and loved?

In fact, if one carefully reads the words and stories and books about Jack Welch, innovation is seldom mentioned. A very well known PhD in electrical engineering and former GE employee said to me at dinner one night, "Welch took a sleepy old electric company and turned it into a bank."

Some truth here. Granted, Jack Welch was immensely successful in turning GE into a bank, and *globalizing* it and *computerizing* it, and improving its quality of product and moving it more into the service business. But he can't really take his place in the world for creating a culture dedicated to innovation.

Marquette Electronics employees say goodbye to Mike after the sale of the company.

CHAPTER 14

THE AFTERMATH

I 'm not sure if anyone has ever had the nerve (or stupidity) to write about the mental anguish a lucky and successful entrepreneur suffers when he's suddenly thrown onto the sidelines after so many years as the boss. My bet is that well-known athletes, musicians, writers, and those in political life could write a thing or two about this subject if they dared.

Nonetheless I will attempt to do so.

I was the boss, no question about it. I started Marquette with one employee and ran the place like a benevolent dictator even when we had thirty-four hundred people. I don't think I ever abused my power, but I always knew it was there. I gave orders, and that's what the troops wanted. I was very lucky in coming up with some innovations for medicine and in so doing, people thought I had some sort of magic. I was working ten, twelve, even fourteen hours a day and I loved every minute. I knew most of our employees by their first names, and they all called me "Mike." Marquette was my home.

When the deal to sell Marquette to GE was negotiated, I was right there across the lunch table with Jack Welch, the then Chairman of the Board, making Irish jokes and saying things were now going to be even *better* for Marquette with GE behind us. Jack said the culture of Marquette would remain forever because Marquette was a great company and, as a division of GE, would stay that way. I felt I had made a very good deal for myself and for my employees.

The luncheon with Jack took place in July of 1998, and in November of that year we signed the papers for almost one billion dollars. Mike's TV repair shop had arrived.

At first, I guess I was numb. I don't think I was fully aware of what was really happening. My left brain knew that I had agreed to sell, but my right brain was unconscious. What had I done? Had I suddenly sold what I had worked so hard for all those thirty-four years, my own child?

Where was my old partner, Warren Cozzens, when I needed him? Unfortunately, he had passed away earlier that year and I never missed him as much as I did at that moment. He probably would have said, "Kee-ryst, Mike, you're seventy-four years old! Go do something else and stop stewing."

Nonetheless, it was all over and now there was only silence. The speeches had all been made, and I was alone with only my ex-wife, Lisa, the poor girl. I sold Marquette and now it was gone. I'd awakened the next day with no regrets, no responsibilities, no cares, no problems – I'd never have to worry about money again.

But wealth was never my goal. What would I do with it, anyway? And what about all those innovations that were still cooking at Marquette? What about all those great people who worked for me, believed in me, trusted me? How would they ever get along without me?

I felt empty. Bewildered. Lost. True, Welch had appointed me as a *special advisor* to GE for four years, but just what did that really mean? Answer – not much.

And then it really hit – the damnedest feeling I have ever experienced. I was down in Florida at the time and suddenly felt sick. Very sick. Physically sick. I was sure I had walking pneumonia. I couldn't breathe! No, it was a cardiac problem. No, that can't be because I was forever testing my heart with our equipment. Perhaps it was something very wrong with my stomach, like cancer. I was dizzy. I must have something wrong with my head, perhaps a tumor of the brain.

I checked myself into the Cleveland Clinic in Ft. Lauderdale and had a blood work-up, an EKG, a urinalysis and a few other tests. The physician in charge called me a few days later and said I had better go and see my urologist right away because I had blood in my urine, and that was a bad sign.

I rushed back to Milwaukee and visited my old friend, Jack Kearns' son, Chris, who had pursued urology just like his old man. Chris said I'd better have a biopsy of my prostate. Of course, then I had that pleasure of waiting, and waiting, and waiting for the test results. Four days of sweating, but all was negative.

Not satisfied, I went out to the Medical College of Wisconsin and had an MRI on my head, a CAT scan on my abdomen and every other test they could think of. All were negative.

A week later I really thought I was going to go mad. I couldn't breathe, damn it. They hadn't found my problem. My ears were ringing! It sounded like I was in a jet, continuously. I was sure to go mad at this rate. I went to see an eye, ear and throat specialist, and he said I had Meniere's syndrome and nothing could be done about it.

Meniere's syndrome. There are three classic symptoms – deafness, head noises, and attacks of vertigo. The cause is unknown, but appears to be related in some cases to Migraine.

The Royal Society of Medicine.

Mike and Lisa

Lisa had to put up with all this nonsense night after night. She'd try to comfort me and tell me to breathe deeply, slowly. I woke up in the night one time and told her I was dying. I said, "Take me into St. Mary's Hospital. I can't breathe. I'm going to croak!" Instead, she gave me a big overdose of Advil. I passed out and woke up the next morning feeling better.

After settling down a bit, I called my old pal and longtime Marquette board member, Walt Robb, in Albany, New York. I wondered how he was doing after his retirement from GE. He was their research director, a big position, indeed, and was required to step down at sixty-five.

"How are ya doing down there, Walt? Getting along OK without all that hustle and bustle of GE Research?"

"Hi, Mike," came an enthusiastic voice at the other end of the line. "Why, let me tell you, I'm busier, and having more fun, than I ever did at GE!"

"Doing what?" I asked, totally surprised.

"Well, I must be on the boards of at least fifteen companies by now, and I've got money in some of them as well. They're all high tech outfits making tremendous strides in things like fiber optics, fuel cells, X-ray technology and even new ways to treat cloth fabrics," he said.

"I'm jealous, Walt. I'm out here wondering what the hell to do with all my energy now that Marquette is gone," I said. "Are you still on the board of that semiconductor company I told you about – Cree, I think the name was?"

"You bet I am. They're doing just great! You still have your stock in the outfit?" knowing full well I sold out awhile back.

"No, I sold it. You know that. How much is it trading for now?" I asked.

Walt chuckled. "Oh, about a hundred. Where did you get out – $28 was it?"

"Don't rub it in, Walt," I said. "Tell me about these other companies you're fooling around with."

"Why don't you come down here, and we'll talk, Mike. You'll be fascinated with all of these companies. I'll introduce you to the key people. They're all right around in this area."

"How about next week, Walt? It shouldn't take but an hour and a half in the jet. Wednesday, OK?"

"Great," said Walt. "I'll pick you up. Get here in time for dinner and then you can stay at my house. Ann would love to see you again."

I hung up and wondered why I had been moping around, feeling sick and sorry for myself all this time since the GE purchase. Perhaps some of these symptoms I was having were psychosomatic. Maybe what I needed was a new purpose in life. Walt was having a ball, surrounding himself with the things he loved and keeping busier than ever.

Soon I was involved in many of the same companies Walt was so excited about. I invested and joined their boards as Walt had and found myself as busy as when I was running Marquette. Most of my *diseases* seemed to vanish as if by magic a short time later.

Besides that, it was time to start thinking seriously about what I was going to do with the fortune I had acquired in GE stock. For the moment, I was GE's biggest private stockholder.

Before the sale of Marquette to GE, I really didn't have much money floating around loose. In spite of my success, most of my money was tied up in the company, and if I went on a stock selling binge, I knew I'd raise all sorts of hell with our stock price. Suspicion would be generated immediately about what might be wrong with the company what with the boss selling out.

I had immense borrowing power at the banks, however, because of my holdings in Marquette, and I had worked my way into a heap of debt by making contributions to local universities and a science museum for kids called Discovery World.

I always had a tendency to give, I guess because I felt I was an unusually lucky fellow. Giving quelled my guilt feelings. This might be uncharacteristic of a guy who loves capitalism and the free enterprise system, but I was always wondering why I had so much and others so little.

And I had plenty of experience with fund raising as well. One of my theories is that an entrepreneur leans towards the generous side. Sort of *easy come, easy go* as a way of life. *Old money*, on the other hand, people who have had money in the family for years, tend to hold on to it for dear life. Fund raising, especially in conservative Milwaukee, Wisconsin, was considered very difficult at best because much of the wealth around was *old money*.

I had lunch with a fellow entrepreneur awhile ago – a guy far more successful than myself, George Dalton. It wasn't supposed to be about any fund raising, but about halfway through lunch, I decided to put the squeeze on him for my favorite cause, the Discovery World Museum for kids. As he finished his wine, he said, "OK Mike, I'll give you $25,000. Now let's talk about *my* favorite charity, The Humane Society. You know we're building a brand new facility out on Wisconsin Avenue."

"All right, George," I said, I'll make a donation, but we could have both stayed at home and given our money to our favorite causes directly, don't you know."

"Ah, you're right, Mike, me boy, but then we wouldn't have had the pleasure of each other's company for lunch," he said in his true Irish style.

I never did subscribe to the idea of giving large amounts of money to my offspring so they could live high on the hog and then pass on what's left to their kids. There's no doubt in my mind that a guy in my position can do more harm than good with such a notion, and fortunately my kids agree.

But having sold Marquette to GE, I felt liberated and set about to make the best of it for the community. I had two wonderful *right hands* that came along with me. Kevin Lindsey, a CPA, a star and with a much more spirited approach than most accountants, left the minute we sold Marquette. JoAnna Hamadi, my secretary/assistant, came along, too. She could keep more things straight than ten other human beings and was charming as well. We formed the Michael J. Cudahy Trust, and we all went to work to see how we could give money, intelligently.

You'd think I would have been pestered to death in such a situation. It was all written in the papers many, many times. How many shares of GE I had and what it all was worth. How I'd said I was going to give a lot of it away. But in Wisconsin, like possibly no other place on earth, most people respected my privacy and left me alone. It seems to me that people from this state are more polite, perhaps more cool about such things, and for that I'm most thankful.

Our main thrust would be the kids in the ghettos. I found out that the U.S. was turning out about a million high school dropouts every year. Europe, on the other hand, was walking all over us with college qualified, terrific young men and women. I felt that if we didn't do something for our kids in the good old USA, we'd really have a grave problem in the years to come.

At this writing, we have granted something more that $50 million to the YMCA, the Boys and Girls Clubs, Marquette University, the Milwaukee School of Engineering, the Medical College of Wisconsin, and a host of smaller schools, "school choice" programs and many more. We believe the kids of this country need all the help they can get and if the money is placed wisely, our foundation *can* make a difference.

We picked organizations like the YMCA because they seemed to be helping solve the problems of inner city kids. So did the Boys and Girls Clubs. They were seriously addressing the never ending cycle of kids with little home life and nothing to do after school but get in trouble. Giving them sports, music, art, computer science and a warm, fuzzy place would keep them off the streets. I gave the old family farm to the "Y" so they could build a new club on the northwest side of Milwaukee. There, the neighborhood had gone to pot with crime and drugs. The "Y" was sure to breathe new life into the area.

I had a ball as this new YMCA started to take shape on the old farm property. At first, it was tough watching bulldozers push around all that land where I grew up. I even found tears in my eyes and had flashbacks of the days when I was a kid there. But when the new facility opened, I was eminently proud and wished my father could see what I was doing.

Some people have had suspicions about my generosity with all this. Why was I giving so much? Was I terminally ill with some dread disease, or was I just plain nuts? Was I trying to make up for all the sins I had committed through the years? No. I asked the good Lord, and he informed me there wasn't that much money in the world. And I'm not trying to show off how successful I've been. As I've said before, I have always felt most of my success was just plain Irish luck.

Sometimes all this giving away money hasn't been easy, however. In one case, donating a small *between acts* bar to the old Pabst Theater turned out to be a nightmare. The trouble started with my old friend, Fred Luber. One day he called and said he had such an opportunity for me to make a contribution that he'd even buy me lunch.

I said, "What's the catch, Fred? You just don't go around spending your money buying people lunch unless you have an evil plot, so what is it?"

"Why would you say a thing like that, Michael? I just wanted to talk to you about the Pabst Theater and the renovation program we're doing there. I'm sure you'd agree that the Pabst is one of our greatest city treasures, wouldn't you?" he said.

"Sure, Fred, but how much are you going to ask me for as you're busy paying the lunch bill?"

"Why, Michael, we're building a bar on the east end of the building, and I thought we could name it "Cudahy's Irish Pub" in your honor since you like to drink so much. It's really a great naming opportunity."

"All right, Fred, how much?"

"Oh, this side of a million. No big deal for you, old buddy," he said.

"Listen, you'll have to buy me more than lunch for that kind of money," I screamed. But I ended up buying lunch and giving in, somehow. I have a weakness for old, classic buildings like the Pabst Theater.

Then the fun started. Fred called about a week before the dedication and said that some of the people from the theater were not all that crazy about the name "Cudahy's Irish Pub," so would I consider changing it to the "Pabst Winter Garden?"

"Are you nuts, Fred? Hell, no! Double hell, no!" I blasted into the phone. "What kind of a naming opportunity was this? Besides, you were the one who suggested the name."

Fred was silent for a few seconds and then quietly assured me that the name would, indeed, be "Cudahy's Irish Pub."

A few days before the dedication, I received another telephone call, this time from the theater president, Dennis Conta. He seemed almost unable to tell me what was on his mind, but finally it developed that he, too, had difficulty with the name.

"Could we just not announce the name at the dedication so we could examine other alternatives?" he asked.

I didn't know Dennis as well as Fred, so I tried to keep my cool, but I was about to lose it. "No! No way, Dennis. Wasn't this a naming opportunity? It sounds to me more like a German conspiracy. First it's Luber. German. Of course, Captain Frederick Pabst, the brewer, was the builder of the damn place about a hundred years ago, and if he wasn't German, who is? Now you call. I'll bet your name used to be Von Conta, and you dropped the *Von*, right?"

"Ah, gee, Mike, don't get your dander up. We'll keep it as *Cudahy's Irish Pub* if you feel that strongly about it."

The next call was from one of the ladies down at the theater. "Oh, Mr. Cudahy, we're so excited! The magicians' convention is in town, and guess what? They want to saw you in half on stage as a part of the dedication. Oh, it's going to be so much fun. The press and the TV will all be there. You don't mind, do you?"

"As long as they can sew me back together again and the name is *Cudahy's Irish Pub*, I guess I'll live through it," I said,

When Irish Eyes Aren't Smiling ...

Donor Cudahy, irked amid naming of Pabst Theater addition, gets his 'pub'

BY MEG KISSINGER
of the *Journal Sentinel* staff

What's in a name? They've done enough Shakespeare in the Pabst Theater to know better than to mess with that question, first posed by the Bard, particularly when the name is Cudahy and he is fixing to donate a pile of cash.

But someone indeed did mess with Cudahy's name, and Michael Cudahy, philanthropist, founder of Marquette Electronics, and namesake for everything from museum wings to university buildings to gardens, is not pleased.

"Oh, they're messing with it, all right," said Cudahy on Thursday, a day after he showed up for the dedication to the new lobby upgrade at the Pabst Theater thinking it would be announced as "Cudahy's Irish Pub" only to find out that there was some confusion over the name. Some folks were calling it a "wintergarden."

"Whatever that is," he sniffed.

Actually, it is neither a garden nor a pub, but a glass-enclosed addition to the lobby, to be constructed on the Water St. side of the theater. The space will allow more room for theatergoers to flow and mingle after shows and during intermissions.

Cudahy, who is giving out more money to Milwaukee lately than Brewers pitchers are giving up runs, says he was approached to be the so-called anchor donor for the $8.8 million theater renovation campaign.

Super Steel Products Chairman Fred Luber ("a German," Cudahy adds) approached him some months ago about the idea of a donation to the theater.

"He told me this would be a great naming opportunity," says Cudahy. "He even suggested the name. He said, 'Cudahy's Irish Pub' had a nice ring to it. And it does."

So, Cudahy signed on the dotted line, pledged $1 million to the campaign.

A few weeks later, Cudahy says, his phone starts to ring.

"Luber's calling me back saying, 'Uh, Mike, uh, uh, some members of the board don't like the name,'" Cudahy says. "I told him, 'Maybe they don't need my money.'"

Next thing he knew, Dennis Conta, chairman of the Pabst Theater Board, was calling.

"Conta. I don't think that's a German name. But it doesn't sound Irish, either," says Cudahy. "I get the same deal from him: 'Mike, uh, uh, uh, about this name…'"

Apparently, says Cudahy, they didn't like the idea of an Irish pub in a building named after a German industrialist.

"Never underestimate an Irishman," says Cudahy. And a stubborn one at that.

Cudahy held firm to his original plan and appeared at the news conference on Wednesday, but there was not so much as a sign acknowledging it as Cudahy's Irish Pub. Augie Pabst, great-grandson of Captain Frederick Pabst, who built the famous theater, was on hand for the big announcement.

"Sounds like a German conspiracy to me," says Cudahy.

Evelyn Woods, campaign manager of the fund drive, had said Wednesday that the name had not yet been decided upon. By Thursday afternoon, she was telling a different tale.

"We are in 100 percent agreement with Michael," she said. "What he wants it to be named, it will be named."

There will be signs in two spots designating it as such, she said.

Luber could not be reached. But Conta, who was on a 100-mile bike trek, called in on his cell phone to say, "the board is prepared to honor Mike's request."

This seemed to please Cudahy.

"*Ach du lieber*, " said he.

wondering if this was another plot to get away from the name I had agreed to.

The whole thing came off without a hitch, however, and I made a little funny speech about the naming opportunity and how it almost went amuck, but now all was well.

Next day it was in the papers – how the new addition, "The Pabst Winter Garden," was a gift from Mike Cudahy – what a nice man.

My blood pressure went through the roof. A few days later, a gossip column picked up on the real story and had a ball with it, as did thousands of readers.

Not much time these days to feel sorry for meself, what with being sawed in two and all. At seventy-seven my only regret is that I'm not forty again, the age when I started Marquette.

The Michael J. Cudahy Foundation
dedicated to philanthropy

Philanthropy, as defined by Webster:
1) Goodwill to fellow men and women
2) An active effort to promote human welfare

Philanthropy, described by Cudahy:
1) The betterment of our children
2) The betterment of our community

In 1964 I, Michael Cudahy, along with my partner, Warren Cozzens, organized a company and called it Marquette Electronics. The objective of the company was to design and manufacture innovative products in the field of medical electronics.

Over the next thirty-five years, Marquette introduced many innovations that improved the practice of medicine and the efficiency of medical care. This was possible through the employment of thousands of talented and dedicated people, many of whom devoted their entire working lives to the company.

In 1998, I sold Marquette to the General Electric Company, benefiting many of these employees as well as the Cozzens family and myself.

Now is the time for philanthropy. Now is the time for me to begin giving back to the community where it all started.

The Michael J. Cudahy Foundation will be guided by its board of directors and invest in a broad selection of securities and businesses with the aim of growing the Foundation as well as giving to the community.

Michael J. Cudahy
January 1, 2001

APPENDIX

In the earlier pages of this book, I referred to my father's fascinating one-on-three interview with Adolph Hitler in 1941 (see page 26). Here is the full text of that interview, as well as some comments by my father on the American press and their handling of the interview.

HITLER ON AMERICA
by John Cudahy

I stood looking for a moment, forgetful of all else, until Schmidt suddenly galvanized to rigid attention and whispered, stagelike, "The Führer."

Hitler stood at the threshold, accompanied by Walter Havel, his liaison officer with the Foreign Office. Schmidt crossed the room and raised his hand on high, to which Hitler replied with the same gesture. Without a word he extended his hand to me and shook hands limply with no display of enthusiasm. I had seen Hitler at Brown Shirt meetings when I visited Berlin during my four years in Poland and only two weeks before in the Reichstag when he addressed that body on the campaign in Greece. In these appearances he had looked so much taller, so much more impressive, that at first I could hardly identify him with the slight figure before me now, whose melancholy fragility reminded me of Harry Hopkins. Above all I was struck by the unhealthy pallor of his skin. He had the same look as prisoners who have been denied the sun during a long period of confinement. He looked as if he might be the victim of a gall bladder, dog-tired, with swollen, puffed eyes, febrilely bright. You could believe the stories that he got less than four hours' rest each night. He gave the impression of being utterly fatigued, one whose nervous energy was nearly spent from overstrain.

Directly after our introduction, Hitler crossed the room and without a word sat down in an easy chair by the round table near the window. A grandfather clock in the corner struck three; Hitler was on time, to the dot. I stood awaiting an invitation from my host to sit down until Schmidt whispered that I should take the seat next to the Führer while Schmidt sat at my right and Havel sat directly opposite Hitler. Hitler stared at me and I stared back. He continued to stare so long that I wondered if this staring duel would ever end. His expression was one of cold hostility. Finally he dropped his eyes and after that only glanced at me casually from time to time. His eyes were truly remarkable and gave the impression of light, so intense they were. They were the arresting feature of his face, harsh, metallic eyes, indicative of an intense, indomitable will, geared to a frenzy. In color they were so pale that at first you could not identify the pigment of the iris. Perhaps they had been seared by gas blindness at Ypres during the last days of fighting in 1918. As our talk developed, I had a chance to examine Hitler's eyes carefully and decided they were that pale translucent green one sees in certain moods of the sea. Above all they were hard, unyielding, fanatical eyes, harsh as the facial lineaments were harsh, without one compromising note of sympathy or kindliness.

As we sat there, Hitler's attitude was that of a man who faced a disagreeable ordeal and wanted to get it over with as soon as possible. He crossed his knees and teetered one foot up and down impatiently. He was dressed in a gray-green coat, like a military tunic, with a double row of brass buttons and wide lapels, wore dark civilian trousers, a white shirt with a soft collar and black tie with a silver swastika scarf pin. After we had talked for a few minutes, he draped one arm about the back of his chair and I had an opportunity to study the hand, a small hand with short square finger tips, a white and lifeless hand of unseen veins like the hand of one dead. All through the conversation he held this hand motionless and never did he gesticulate with either hand.

There is more in any talk than what is spoken. A test of this is to put a wall of paper between you and the person with whom you are talking. Instantly you become detached, that strange magnetism of visual human presence and expression, and the connotation of the unspoken word taken away completely. The same thing in some measure is true when speaking through an interpreter, for it seems silly to look at one and address one directly in a language which is meaningless as Chinese or pidgin jargon. So in this conversation my questions were directed to the interpreter, Schmidt, but Hitler strangely when he answered, spoke to Havel, an old comrade of the first Putsch days, and seemed to find a response and receptiveness in him, rather than in the interpreter.

No written questions were submitted before the interview and Hitler had no intimation of what I would ask him. The issue of convoys was agitating the American people and the ignorance on this subject was unbelievable. We Americans are a land people and know little about nautical terms or anything touching upon the sea. Many persons in states like my own Wisconsin had no conception of what was meant by a convoy or its significance, so I considered the most important issue to be clarified was this matter of convoys and Hitler's attitude if American ships should escort war materials to Britain. He told me very simply and with no show or dramatics in answer to this question, that convoys meant war and went on to say that the action of escorting munitions and deadly weapons to an enemy with armed naval forces had always been considered a warlike act, the

The Great Hall of Berghof was the scene of Mr. Cudahy's interview with Adolf Hitler. The great window at one end used to be in sections but Hitler had it made a single sheet of glass to afford a better view of his beloved, snow-capped mountains. Of this room Mr. Cudahy cables, "I was met in the hallway by Walter Havel and a captain aide. I distinguished a portrait of Bismarck as we went down a passageway and through doors to an oblong room of great height, length and breadth. We descended three steps. At the opposite end of the hall another stairway with iron

legal precedents had been determined by Anglo-Saxon maritime powers and were thoroughly well known and understood by all legalistic authorities the world over.

I had always been told that Hitler's mental processes were emotional rather than intellectual, but that was not my experience at this meeting. He spoke concisely, consecutively, and more responsibly than many men of state I had

balustrade leads to the only other exit, a Roman-arched doorway. The whole color scheme has a garnet tint: the carpet, the marble steps and the covering of the furniture. On both white plaster walls there are swastikas, tapestries and paintings of reclining nudes. The woodwork and the paneling on the ceiling are of shellacked oak. I noticed an oak table, a piano and a bust of Wagner. There were calla lilies and carnations on a table, and hydrangeas in a bowl. A clock struck noisily during our conversation."

interviewed under the cloak of confidence and secrecy when I was in the Foreign Service. His voice had the harsh, frayed quality one associates with political orators at the close of a hard campaign and was utterly lacking in any sympathetic timbre. I did not find Hitler a monologist or given to a great volume of words. So many people had told me about his ranting and raving, yet in this conference his voice was never once raised and never did he give sign of any agitation; nor did he gesticulate, but spoke with the utmost composure, never betraying vocally the intensity indicated by the taut lineaments of his face.

His hair was a plastered mouse-brown mop. His mustache showed a few gray hairs and there was a hint of gray commencing at the temple and back of the ears. The forehead showed a remarkable protuberance above the eyebrows, which the phrenologists call the perceptive cranial area. The upper forehead receded and did not indicate great contemplative capacity. The nose was thick and heavy, without clean-cut line, and the lower face, although not heavily boned or projected, gave an impression of great power of will, great energy, and aggressiveness.

I told Hitler that the reason the United States had passed from stage after stage of hostility to one of non-combative belligerency was because many Americans feared first the probability of invasion and next the competition of German goods in world commercial markets. These were the primary reasons for American hostility to Germany in the war, I said, and referring to the first of these I said there was widespread sentiment among American people that the security of the Western Hemisphere was threatened by German aggression. It was argued that the German conquest might go on and on and with Europe conquered, the next logical step for the German military adventure was the two American continents. He laughed a harsh, strident laugh, disagreeable as a rasping automobile gear and his face looked as if it had taken a long holiday from honest, spontaneous laughter when he said that the idea of Germany invading the Western Hemisphere was about as fantastic as an invasion of the moon. I replied that fantastic or not, an eventual attack by Germany on the Americas was feared by a large number of thoughtful American people. He said he could not believe this, he had too high an opinion of American intelligence and good sense. He turned on me sharply and asked the opinion of American military experts on the subject. I told him I did not know, but believed that the Chief of Staff and Chief of Naval Operations had never been quoted. He went on then to say he was convinced that if they were ever heard from they would confirm what he had first said. This invasion story was, he was certain, put out by warmongers against their better knowledge, men who wanted war in the belief that it would be profitable for business, but this was an erroneous conception since the last great war had demonstrated war was ruinous to business.

He declared that the German High Command regarded an invasion of either American continent as wildly imaginary. He asked why the British did not send more troops to Greece and North Africa in the campaigns in these countries and answered the question by saying it was because sufficient transports were not available, although the distances were comparatively short. He assured me that the combined shipping tonnage of the British, United States and German marine would likewise be hopelessly inadequate to transport the army of millions which would be required for a successful conquest of the Western Hemisphere. The German army, he assured me, was not concerned with military expeditions for the sake of showing off or in order to demonstrate that nothing was impossible for German arms. The undertaking of invading Crete over one hundred kilometers of open water and England over forty kilometers of open water were formidable enough, and he said the thought of an attack over the four thousand kilometers of open water between Europe and the United States was simply unthinkable.

Somewhat irrelevantly he then declared that nobody in his hearing had ever said that the Mississippi was the German frontier in the same spirit that the Prime Minister of Australia had referred to the Rhine as the frontier of that country, but since the Rhine was their frontier, he had decided to send some Australian prisoners to the famous German river so that they might acquaint themselves with the frontier atmosphere. He assured me that Germany had too many serious problems in Europe ever to give any thought to an American invasion.

Many people, I told the German Führer, shared his view that military science had not yet developed to such a point that the Atlantic could be brushed aside as an obstacle of defense for the Western Hemisphere, but many people believed that the greatest menace of a triumphant Germany was to the economic life of the United States, and that German victory would mean disaster to American business. This conviction, I said, was predicated upon the assumption that the standard of living in Germany was far lower than that in the United States. Also the working hours imposed and the regimentation of German labor would never be accepted by working people in the United States. Therefore, it was only logical to assume that American industrial output could not compete with that of Germany. He did not like the inference that the living standard of the German worker was a low one, and said the controlling purpose of National Socialism from the beginning had been to improve living conditions for labor. The war had interrupted his effort, but it would be renewed with redoubled force when peace came. He said one of the controlling ambitions of his life was to improve the lot of the common man in Germany and among other things he hoped to see every laboring man own an automobile. He reminded me then that Germany with a population density of 140 per square kilometer had conquered the depression and provided jobs for all so that there was no longer any unemployment in the country, while the United States with eleven to the square kilometer was unable to cope with its serious unemployment problem. Why, he asked me, did we single out Germany as the outstanding menace to American economy, when Germany had a physical area of six to seven million square kilometers, and a population of eighty-five million, while the British Empire comprised four hundred million kilometers, Japan one hundred million kilometers, Russia one hundred seventy million, and other nations of the world four hundred million. These national areas, he intimated, were far more menacing to American economy than Germany was. He asked why if German competition was so greatly feared, were her colonies taken away from her, and went on obliquely to say that development of these colonies would have provided a great outlet for German industrial output.

Why, he asked me, was the United States opposed to German organization of Europe in order to provide a market for German goods, thereby lessening competition with the United States in the world markets? He talked at some length of southeastern Europe, which he said was the natural complement to the German economy, because the Balkan countries had surplus agricultural produce, which they could exchange for German industrial output. He declared there was an *iron rule of trade* which was that no country could buy from another unless it could also sell. Assuming the verity of this rule, he asked how the United States, with great agricultural surpluses, could take further farm produce from southeastern Europe, which was the only payment that area could make for needed American manufactured articles. I inquired whether or not he envisaged a trade union for Europe with suppression of quotas, tariffs, currency restrictions, etc. He replied that he thought all commercial relations between all countries could only be assured by long-term trade treaties permitting the partners there to a profitable arrangement and suppressing the element of speculation, which he said had always cursed business. He did not believe in trade relations based on borrowed money and declared loans have to be paid and the end of borrowing often is disillusion and bankruptcy. The future trade of Germany, he insisted vigorously, would not be based on paper but upon exchange of commodity for commodity with an absolute exclusion of all speculation. The professors, he said sneeringly, had scorned these economic theories of his, but in twenty or thirty years, he predicted, they would be teaching them in the universities. He was frank in saying he saw no prosperous future in the trade relations of the

United States and Germany for the reason that Germany had no gold with which to buy the raw materials which it would need and the United States would not take in payment German manufactured goods to compete with its own output, except perhaps in a limited field of electrical equipment, optical goods and possibly dyes.

I asked about gold and its function in the future international trade of Germany. He said that reparations had deprived Germany of all its gold and had forced him to devise a system of international trade without gold, but I was surprised when upon further questioning, he admitted the usefulness of gold in providing a more elastic method of mercantile dealing and as a basis of credit between nations. He said quite openly that Germany was forced to trade by barter and had been driven away by necessity from the use of gold.

I then turned to the countries occupied by German military forces and, although I had been warned that the Führer would not discuss this subject, I pressed him to indicate his thought with reference to such nations, telling him frankly that my question was inspired by a violent prejudice among Americans that German domination of Europe would mean suppression of native national language, customs, and institutions in the occupied territories. His unresponsive answer was that Germany had not commenced this war, the war had been declared by England and France against Germany and it was strange to hear the British discourse on world domination when they went on suppressing millions of Indians, Egyptians and Arabs. "We shall settle relations with our neighbors in such a way that all will enjoy peace and prosperity," he summarized. I became more specific and referred to the case of Belgium, explaining that my interest had a personal angle because I had lived in the country. His answer was that his formula for the future of Europe was "Peace, prosperity and happiness." Germany, he said, was not interested in slaves or the enslavement of any people.

As we talked on, the clock in the corner struck at several intervals with rasping regularity. And during one of these clanging outbursts, the Führer rose to his feet in the same abrupt, ill-mannered fashion in which he had sat down. He shook hands with me with the same unmollified hostility he had displayed at the outset, saying that he had tried to answer all my inquiries with clarity and without reservation but he believed the interview was a waste of time, since no matter what he said his words would be distorted by the American press. I told him I would report what he said honestly and objectively without comment, but I had no idea of how American editors might construe and interpret his remarks. We mounted the marble steps and at the threshold of the Berghof, Schmidt turned and raised his hand ceremonially in farewell. Then we went away.

This interview was submitted to Hitler and approved by him without change. It was then cabled to *Life* and Time Publishing Company and published a week later simultaneously by *Life* magazine and the newspapers in the syndicate North American News Alliance. *The New York Times* printed it with an editorial disparaging Hitler's words and pointing to his many broken promises. *Life* also published a caption making its position clear that it did not subscribe to anything Hitler said.

The interview covered three salient points, of which the only important one dealt with the question of convoys. On this Hitler spoke succinctly and unequivocally, in saying that the convoy of war materials to England by armed American naval vessels would mean war. It was strange not a single editor or commentator passed upon this, although it was the only significance of what Hitler said to me.

The second subject concerned invasion of the Western Hemisphere. Opinions on this differ, and a very useful purpose would have been served if American experts had replied to Hitler's statement that the German military establishment had decided that an invasion of the Western Hemisphere was not practicable. Instead American journalistic opinion and politicians all declared Hitler had said that he had no designs on the Americas. Editorial after editorial ridiculed and derided such a hypocritical, lamblike attitude and held up his shameless record in Poland, the Low Countries, Austria, Czechoslovakia, etc., to prove that under no circumstances could his word be trusted. But Hitler never said anything about his intentions. His answers to questions dealing with the Western Hemisphere were confined solely to the military aspect of the situation and mentioned nothing else.

This phase of Hitler's utterance absorbed editorial and political attention so exclusively that nothing was said about the third feature, German-American competition in world markets after the war. On this also opinions differ, and a number of very competent businessmen in this country believe that American industry can never compete with German industrial output because of lower German labor costs and the barter system of German international trade.

Mr. Bernard Baruch takes the opposite view and, in a discussion with him a few weeks after my return, he said most emphatically that he believed that American mass production could easily undersell German manufactured goods in any world market. The subject is a very important, far-reaching one and should be discussed and analyzed on its merits without prejudice.

Unlike *Life* and *The Times*, editors, commentators, columnists, and politicians did not confine their criticism to Hitler's words. They directed their remarks to the interviewer, stating openly in some cases and intimating by broad inference in others, that he had been influenced – even bribed – by the Nazis. Others accused him of dopily acting as a mouthpiece for Nazi propaganda and deplored his gullibility, guilelessness, and naïveté. The import of the whole assault was that it was the duty of the reporter to evaluate and comment upon Hitler's utterance instead of recording factually what was said and letting the words speak for themselves. Almost without exception all reviewers did not analyze what was said. They drew conclusions and made assumptions from Hitler's words. This is a familiar practice with which every trial lawyer is familiar and always occurs among ignorant witnesses and those of little mental discipline. Prejudice could have the same blinding influence as ignorance, it appeared. Many prominent educators insist there is no such thing as the judicial faculty or the objective mind. I had a feeling of profound depression, not because of barbs thrown at me personally, but at this exposure of our intellectual level.

INDEX

NOTE: *Italic* page numbers refer to photos and their captions.